Lecture Notes in Mathematics

Editors:
A. Dold, Heidelberg
F. Takens, Groningen
B. Teissier, Paris

Springer

Berlin
Heidelberg
New York
Barcelona
Budapest
Hong Kong
London
Milan
Paris
Singapore
Tokyo

William Fulton   Piotr Pragacz

Schubert Varieties
and Degeneracy Loci

 Springer

Authors

William Fulton
Department of Mathematics
University of Michigan
525 E. University
Ann Arbor, MI 48109-1109, USA
e-mail: fulton@math.lsa.umich.edu

Piotr Pragacz
Institute of Mathematics
Polish Academy of Sciences
Chopina 12
87-100 Torun, Poland
e-mail: pragacz@mat.uni.torun.pl

Cataloging-in-Publication Data applied for

Die Deutsche Bibliothek - CIP-Einheitsaufnahme

Fulton, William:
Schubert varieties and degeneracy loci / William Fulton/Piotr
Pragacz. - Berlin ; Heidelberg ; New York ; Barcelona ; Budapest ;
Hong Kong ; London ; Milan ; Paris ; Santa Clara ; Singapore ;
Tokyo : Springer, 1998
   (Lecture notes in mathematics ; 1689)
   ISBN 3-540-64538-1

Mathematics Subject Classification (1991):
14M15, 14C17, 14M12, 14F05, 14C15

ISSN 0075-8434
ISBN 3-540-64538-1 Springer-Verlag Berlin Heidelberg New York

Typesetting: by the authors using A_MS-T_EX
SPIN: 10649864   46/3143-543210 - Printed on acid-free paper

To the memory of Michael Schneider

PREFACE

These notes grew from lectures given in a Summer School at Thurnau, Germany, June 19-23, 1995.[1] They are designed to give an introduction to Schubert varieties and related degeneracy loci. The subject has its origins in classical questions about loci of matrices of various ranks, as well as the Schubert calculus of intersection theory on Grassmannians and flag varieties. We describe these origins, as well as modern formulations involving vector bundles on arbitrary algebraic varieties.

Most of what is written here describes what was known in 1995; a few remarks, footnotes, and some of the appendices indicate some closely related work that has been done since then.

Chapter 1 describes the classical problems about loci where matrices have a given rank that gave rise to these problems, and states classical and modern solutions, which involve the Schur polynomials and Schubert polynomials that will appear frequently in these lectures. Chapter 2 gives the modern generalization to maps of vector bundles, where the problem is to find a formula for the cohomology class of a degeneracy locus in terms of Chern classes of the bundles. Here we discuss the fundamental case of Grassmannians and flag manifolds, and sketch the proof, which is based on correspondences.

In Chapter 3 the basic symmetric functions that appear in this story — Schur S-polynomials, Q-polynomials and supersymmetric polynomials — are described, with many of their basic properties. In Chapter 4, a useful general technique of Gysin maps is discussed for Grassmannian and flag bundles. These two chapters also discuss the problem of finding which polynomials are universally supported on degeneracy loci with an explicit description of the ideal of all such polynomials associated with the rth degeneracy locus. Chapter 5 contains an application of the technique of polynomials universally supported on degeneracy loci to computation of the topological Euler characteristic of degeneracy loci and Brill-Noether loci in Jacobians.

In Chapter 6 we describe the flag varieties for the other classical groups, and discuss the corresponding degeneracy loci for the generalizations to vector bundles, following the correspondence method of Chapter 3. The approach to these problems via Gysin maps, diagonals and symmetric $\widetilde{P}$- and $\widetilde{Q}$-polynomials is given in Chapter 7.

---

[1] The session also included talks of S. Billey, D. Edidin, W. Graham, M. Haiman, L. Manivel, and J. Ratajski.

One important application of this, to Prym varieties, using the case of the even orthogonal group, is given in some detail in Chapter 8. Several applications and open problems are mentioned in Chapters 1-8. The final Chapter 9 gives a variety of other applications and variations, and lists some other open problems.

The book can be read at several levels, depending on the background of the reader in intersection theory and/or topology. We need at least the basic result that a subvariety V of a nonsingular variety X determines a class, denoted $[V]$, either in a homology group or a Chow group.[2] We use a few basic facts about the relation of cohomology rings with intersections of varieties, and about Chern classes; for a brief discussion of this, see Appendix A.

In the text we will usually assume that the loci we are talking about have the expected codimension, and that the ambient variety is nonsingular. Appendix A gives a general discussion of how the formulas are to be interpreted in more general situations.

There are a few results here that are new. One, given in Appendix B and Chapter 2, improves the main theorem of [F2] to allow degeneracy loci given by rank conditions that may be redundant. This gives alternative determinantal expressions for the loci described in [K-L], [Las3], [P1], [P4] and [F2].

Although the resolutions of singularities described by Demazure [D2] are not needed in our approaches, they are closely related, and described in Appendix C. The definition and basic properties of Pfaffians are described in Appendix D.

The constructions and proofs in the text are largely independent of group theory. Several approaches that bring out relations with group theory are discussed in Appendix E. Appendix F proves a useful Gysin formula for Grassmann bundles.

Appendix G gives a general criterion, somewhat stronger than that in [P4], for the computation of the class of a relative diagonal. A construction of Mumford [Mu1] that is needed in Chapter 8 is described in Appendix H.

We do not discuss any related representation theory, but we do include a simple construction of the irreducible polynomial representations of the general linear groups in Appendix I.

With the advent of quantum cohomology, much of the classical story about flag manifolds is being revitalized with a quantum analogue. The final Appendix J, which is an expansion of [CF-F], gives a very brief description of quantum Schubert polynomials.

Except for the added appendices, the text follows closely the lectures given at Thurnau. We include examples and indications of proofs, but often refer to the original papers for technical details.

Although we sketch some of the history, the bibliography is *not* meant to be a guide to the literature. Rather, we list those papers and books where the reader can find facts we need in the text, or where the latest version of a theorem can be found.

---

[2]For purposes of these lectures, it does not matter much which interpretation is used. We generally use the topological notation for homology and cohomology, since that may be familiar to more readers, but, unless explicitly stated, the results and proofs are the same for the Chow groups.

A reader who wants to find the original sources should consult the bibliographies of these articles.

This is neither a textbook, nor a research monograph, nor a survey. Instead, the authors have tried to describe what they regard as essential features of the story, each from his own point of view. Chapters 1, 2, 6, and 9 correspond to lectures of the first author, and Chapters 3, 4, 5, 7, and 8 to those of the second. We hope that, despite differing styles and viewpoints, a coherent picture of this fascinating subject emerges.

The first author thanks the Mathematics Department of the University of Pennsylvania, where a first version to his lectures were given as the Rademacher Lectures in 1995. The second author thanks for hospitality the University of Chicago where a significant part of this book was written in May 1996. We thank Ionuţ Ciocan-Fontanine for permission to include the preprint [CF-F] in this book. We are grateful to A. Buch, G. van der Geer, W. Graham and A. Langer for comments which led to improvements of the text. Finally, we thank Jan K. Kowalski for his help in preparing some figures in $\mathcal{AMS}$-TEX.[3]

We dedicate this book to the memory of Michael Schneider, who died in a climbing accident on August 29, 1997. It was he who planned and organized the Summer School in Thurnau. Michael was a good friend to us and a good friend to mathematics.

---

[3]The authors also acknowledge research support from NSF grant DMS-9600059 and KBN grant 2P03A 05112, respectively.

CONTENTS

CHAPTER I

INTRODUCTION TO DEGENERACY LOCI AND SCHUBERT POLYNOMIALS

The aim of this first lecture is to give a leisurely, semi-historical introduction to some of the main ideas we will see. There will be some skipping back and forth, however, in order to introduce early some of the ideas that will be featured prominently in this book.

Section 1.1 Early history

This story can be regarded as a continuation of what is now known as Bézout's theorem, which dates back more than two centuries:

> The number of common solutions of m polynomials in m variables is the product of the degrees of the equations.

This had been pointed out and used also by many others such as Newton, and in some ways can be regarded as the origin of algebraic geometry as a subject. Some comments should perhaps be made:

(1) Complex solutions, or solutions in an algebraically closed field, must be included, even if one starts with polynomials over smaller fields such as $\mathbb{Q}$ or $\mathbb{R}$.

(2) One should compactify, homogenizing the polynomials, and considering solutions not in affine space $\mathbb{A}^n = \mathbb{C}^n$ but in projective space $\mathbb{P}^n$. (Note that if polynomials are sufficiently general, however, all the solutions will be in the affine part.)

(3) One can relax further, as long as there are a finite number of solutions, and assign a multiplicity to each solution, so that the sum of the multiplicities is the product of the degrees.

Bézout's theorem can be justified, as it was classically, by the principle of continuity (Poncelet, 1822): the answer doesn't vary as the hypersurfaces vary, and one can deform to a situation one understands, for, example, to the case where each is a product of linear factors, and all the linear factors meet transversally, or even to multiples of linear factors.

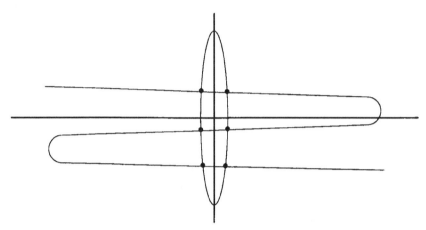

One way to justify this today is to use the cohomology classes defined by the hypersurfaces, and the fact that the product of these classes is represented by their geometric intersection. The class $[H]$ of a hypersurface H of degree d is d times the class h of a hyperplane, and $H^*(\mathbb{P}^m) = \mathbb{Z}[h]/(h^{m+1})$, with h^m representing a point. This leads to the equation $[H_1] \cdot \ldots \cdot [H_m] = d_1 \cdot \ldots \cdot d_m \, h^m$; if the hypersurfaces meet transversally, this implies that their intersection $H_1 \cap \cdots \cap H_m$ consists of $d_1 \cdot \ldots \cdot d_m$ distinct points. Or one can use intersection theory, with Chow groups, to do the same thing.

More generally, consider forms (homogeneous polynomials) $f_1, \ldots, f_m$ of degrees $d_1, \ldots, d_m$ in $N+1$ variables $x_0, \ldots, x_N$, and let $V = V(f_1, \ldots, f_m) \subset \mathbb{P}^N$ be the locus of zeros. Assume $N \geq m$. We want to describe V. First, its expected codimension is m. Every irreducible component has codimension at most m. If the forms are general, and $N > m$, then V is in fact an irreducible subvariety of codimension m. The simplest invariant of a subvariety is its degree: the number of intersection points with a general linear subspace of complementary dimension. Here Bézout's theorem says that for general forms, the degree of V is $d_1 \cdot \ldots \cdot d_m$. By intersecting with $N - m$ general linear forms, the question is reduced to the case of points.

At least until later, we will assume, often without much comment, that forms are "generic" or "general" in some sense. Modern intersection theory allows considerable relaxation of these hypotheses, but it would clutter the discussion to add these hypotheses at each stage. See Appendix A for a discussion of this.

It may be worth pointing out that Bézout's book is mainly a discussion of how to solve equations, i.e., "elimination theory." On one page he does point out that three quadric surfaces in space intersect in 8 points, but that is the only mention of geometry!

The story continues in the 19th century, when mathematicians came up against the following problem, usually arising from a problem in geometry:

We are given an $l \times m$ matrix $A = (a_{i,j})$ of forms in variables $x_0, \ldots, x_N$. Let $0 \leq r < \min(l, m)$. Consider the locus V_r of points in $\mathbb{P}^N$ at which the rank of A is at most r. This will be defined (cut out) by the minors of size $r + 1$ of A. The problem is to describe V_r.

When $r = 0$, this is just the zeros of all the entries, and the answer is given by Bézout's theorem. For $r > 0$, in order for this problem to make sense, these minors must be homogeneous, which need not be the case if the degrees of the entries are arbitrary. To assure that all minors are homogeneous, they assumed (and we assume) that

$$\text{degree}(a_{i,j}) = p_i + q_j$$

for some integers $p_1, \ldots, p_l, q_1, \ldots, q_m$. One first has some preliminary questions: What is the codimension of V_r? Is it irreducible? In fact, the codimension of V_r is $(l - r)(m - r)$, and it is irreducible if this number is less than N, and the degrees of the (generic) forms are all positive. We will see a reason for this in the next lecture. We will always assume as part of "general" that the expected codimension is at most the ambient dimension. The fundamental question is:

What is the degree of V_r?

The difficulty is that there are usually far too many equations — one for each such minor, so $\binom{l}{r+1}\binom{m}{r+1}$ in all. Thus this and related questions became a kind of test of intersection theory in the 19th century.

Cayley, in 1849, seems to have been the first to raise this question, in the case when all the entries had the same degree, and he solved some cases. This was carried on by Salmon and by S. Roberts and others in the 1850's and 1860's. They were concerned with the locus where the matrix fails to have maximal rank, and they gave a complete formula in this case. Of course, if $l = m$, the codimension is 1, and it is defined by the determinant, whose degree is $\sum_{i=1}^{m}(p_i + q_i)$.

The first interesting case is that of 2×3 matrices. The codimension in this case is $(2 - 1)(3 - 1) = 2$, but the locus is defined by three equations — the three 2×2 minors of the matrix. Let us look at a concrete example, where

$$A = \begin{bmatrix} x_0 & x_1 & x_2 \\ x_3 & x_4 & x_5 \end{bmatrix}.$$

We are looking at the locus V in $\mathbb{P}^5$ where the rank of A is 1, defined by the three quadratic equations $x_0 x_4 - x_3 x_1 = 0$, $x_0 x_5 - x_3 x_2 = 0$, and $x_1 x_5 - x_4 x_2 = 0$. In fact, V is the image of $\mathbb{P}^2 \times \mathbb{P}^1$ by the Segre embedding:

$$[s_0 : s_1 : s_2] \times [t_0 : t_1] \mapsto [s_0 t_0 : s_1 t_0 : s_2 t_0 : s_0 t_1 : s_1 t_1 : s_2 t_1].$$

The degree of this subvariety is 3. One can see this by noting that the hyperplane section intersects this subvariety in a class equivalent to $h_1 + h_2$, where h_1 and h_2 are classes of hyperplanes in $\mathbb{P}^2$ and $\mathbb{P}^1$ respectively. So the degree is the computed by calculating (in the cohomology or Chow ring):

$$(h_1 + h_2)^3 = 3\, h_1^2\, h_2 = 3\,[\text{point}].$$

Note that this implies (by Bézout) that V cannot be cut out by two equations, since 3 is prime, and V is not contained in a hyperplane.

EXERCISE. Show that the Segre embedding of $\mathbb{P}^{m-1} \times \mathbb{P}^1$ in $\mathbb{P}^{2m-1}$ has image the locus where a $2 \times m$ matrix of linear forms has rank 1, a locus defined by $\binom{m}{2}$ quadratic equations. Show that the degree of this locus is m.

In general, for the 2×3 case, the degree of $V = V_1$ was found to be

$$(1.1) \qquad p_1^2 + p_1 \, p_2 + p_2^2 + (p_1 + p_2)(q_1 + q_2 + q_3) + q_1 \, q_2 + q_1 \, q_3 + q_2 \, q_3.$$

One can take all $p_i = 1$ and all $q_j = 0$ (or one can take all $p_i = 0$ and all $q_j = 1$), and recover the degree of 3 for the above example. It is interesting to explain the proof that was given in this case. Consider the intersection of the two hypersurfaces defined by two of the minors, $a_{1,1} \cdot a_{2,2} - a_{2,1} \cdot a_{1,2}$ and $a_{1,1} \cdot a_{2,3} - a_{2,1} \cdot a_{1,3}$, which intersect, by Bézout, in a subvariety of codimension two and degree

$$(p_1 + p_2 + q_1 + q_2)(p_1 + p_2 + q_1 + q_3).$$

Now this intersection is the union of the locus V_1 that we want, together with the locus defined by the equations $a_{1,1} = 0$ and $a_{2,1} = 0$. Since the last locus has (by Bézout) degree $(p_1 + q_1)(p_2 + q_1)$, we get

$$(p_1 + p_2 + q_1 + q_2)(p_1 + p_2 + q_1 + q_3) = \deg(V_1) + (p_1 + q_1)(p_2 + q_1),$$

which is equivalent to (1.1).

Here is simple geometric problem of the sort considered then. Consider three pencils of plane curves, of the form

$$F + tF' = 0, \quad G + tG' = 0, \quad \text{and} \quad H + tH' = 0,$$

where F and F' are forms of degree f, G and G' are forms of degree g, and H and H' are forms of degree h. For how many numbers t do the three curves have a common point of intersection? The points of intersection are the points where the matrix

$$\begin{bmatrix} F & G & H \\ F' & G' & H' \end{bmatrix}$$

drops rank, and it has the vector $(1, t)$ in its kernel (with the convention here that matrices act on the right on row spaces). By (1.1), the number of such points, and the number of t, if the forms are sufficiently general, is $f\,g + f\,h + g\,h$.

Salmon and Roberts gave the complete general answer for the degree of the locus where a matrix is singular. Suppose (with no loss of generality) that $l \leq m$, so $r = l - 1$. The codimension is $m - l + 1$. Let h_i be the ith complete (homogeneous) symmetric polynomial in the variables $p_1, \ldots, p_l$, and let e_j be the jth elementary symmetric polynomial in $q_1, \ldots, q_m$. Their formula reads:

$$(1.2) \qquad \deg(V_r) = h_{m-l+1} + h_{m-l} \cdot e_1 + h_{m-l-1} \cdot e_2 + \cdots + e_{m-l+1}.$$

Equivalently, define integers c_k by the identity

$$(1.3) \qquad \sum_{k \geq 0} c_k \, T^k = \prod_{j=1}^{m} (1 + q_j T) \Big/ \prod_{i=1}^{l} (1 - p_i T).$$

i.e., $c_k = h_k + h_{k-1} e_1 + \cdots + h_1 e_{k-1} + e_k$. Then $\deg(V_r) = c_{m-l+1}$.

Tracing the history of this problem during the last half of the 19th century would be an interesting project, especially if one tried to determine what was actually proved. (A paper by E. J. Nanson in 1903 was written for the stated reason that formula (1.2) had never been proved, but it contains scant reference to the literature.) In such an investigation one would meet many well known geometers from this period, such as Cremona, Veronese, Schubert, Segre, Pieri, and Giambelli, among many others. There were several different but equivalent forms of some answers, the equivalence amounting to interesting identities among symmetric functions. Some of the nicest expressions for these degrees were in terms of determinants.

A general formula for the degree of V_r is, with the c_k as defined in (1.3),

$$(1.4) \qquad \deg(V_r) = \det (c_{m-r+j-i})_{1 \le i,j \le l-r}.$$

Giambelli (1903-4) gave this formula, at least if all $p_i = 0$, so $c_k = e_k$. He attributed this case to Segre (1900), coming from a formula Schubert had proved (1894-5) about the Grassmannian.[1]

In this story, determinants like this appear often. They are often called **Schur determinants**. There is one for any sequence $\lambda = (\lambda_1, \lambda_2, \ldots \lambda_n)$ of integers, and any sequence $c_0, c_1, \ldots$ of commuting elements in a ring. This is often written as a formal series $c = c_0 + c_1 + \ldots$. Usually λ is a **partition**, i.e., $\lambda_1 \ge \lambda_2 \ge \cdots \ge \lambda_n \ge 0$, but occasionally the general case is needed. Usually one also has $c_0 = 1$, but we will in one instance need the case when $c_0 = 2$. In any case, one sets $c_k = 0$ for $k < 0$. Define $\Delta_\lambda(c) = \det (c_{\lambda_i + j - i})_{1 \le i,j \le n}$, i.e.,

$$(1.5) \qquad \Delta_\lambda(c) = \begin{vmatrix} c_{\lambda_1} & c_{\lambda_1 + 1} & \cdots & c_{\lambda_1 + n - 1} \\ c_{\lambda_2 - 1} & c_{\lambda_2} & \cdots & c_{\lambda_2 + n - 2} \\ \vdots & \vdots & \vdots & \vdots \\ c_{\lambda_n - n + 1} & \cdots & c_{\lambda_n - 1} & c_{\lambda_n} \end{vmatrix}$$

For example, $\Delta_{(n)}(c) = c_n$, and $\Delta_{(1,1)}(c) = c_1{}^2 - c_0 c_2$, $\Delta_{(2,1)}(c) = c_1 c_2 - c_0 c_3$, and $\Delta_{(1,1,1)}(c) = c_1{}^3 - 2c_1 c_2 + c_0 c_3$. When $c_0 = 1$, the determinant is unchanged if zeros are added to the sequence λ. When c_k is the kth complete symmetric polynomial in some variables $x_1, \ldots, x_l$, this is the **Schur polynomial**:

$$(1.6) \qquad \Delta_\lambda(c) = s_\lambda(x_1, \ldots, x_l).$$

More generally, when c_k is defined by formula (1.3), then $\Delta_\lambda(c)$ is a symmetric polynomial known variously as a **hook Schur polynomial, supersymmetric**

---

[1]In fact, we have not found exactly this general formula in the classical literature. One form of Giambelli's general formula is the determinant of a matrix of size $l + m - 2r$. The first $l - r$ entries in the ith column are $(-1)^{i-1} e_{i-1}, (-1)^{i-2} e_{i-2}, \ldots, (-1)^{i-l+r} e_{i-l+r}$, and the last $m - r$ entries in the ith column are $d_{i-1}, d_{i-2}, \ldots, d_{i-m+r}$, where d_j is the jth elementary symmetric function in the variables $p_1, \ldots, p_l$. It is not hard to show that this determant is equal to that in (1.4).

Schur polynomial or **bisymmetric polynomial**. These and other symmetric polynomials will be discussed in Chapter 3.

The formula for the degree of V_r reads: $\deg(V_r) = \Delta_\lambda(c)$ where $\lambda = ((m-r)^{l-r})$ is the partition consisting of $l - r$ copies of $m - r$, and c_k is defined in (1.3).

Note that the expression (1.4) changes completely if one considers the transpose of A, although the locus is unchanged. This amounts to a well-known duality formula:

$$(1.7) \qquad\qquad \Delta_{\tilde\lambda}(\tilde c) = \Delta_\lambda(c),$$

where $\tilde\lambda$ is the conjugate partition of λ (obtained by interchanging rows and columns in the corresponding Young or Ferrers diagram), and the $\tilde c_k$ are determined by the identity $\left(\sum(-1)^i \tilde c_i T^i\right) \cdot \left(\sum c_j T^j\right) = 1$. Giambelli also gave this dual formulation of (1.4), in the case when all $q_j = 0$, so $\tilde c_k$ is the kth elementary symmetric polynomial in the variables $p_1, \ldots, p_l$.

There are many other related loci that were studied, particularly by Giambelli. In addition to requiring that the rank of A be at most r, one can, for integers s and t between 0 and r, look at the locus where the rank of A is at most r, and the top $s \times m$ submatrix of A is singular (has rank strictly less than s), and the left $l \times t$ submatrix is singular (of rank $< t$). Giambelli gave a general solution to this, for arbitrary p_i's and q_j's, as a sum of "Jacobi-Trudi" ratios.

Giambelli also gave formulas for a collection of top and left full submatrices to have various ranks. Here, however, he had to make assumptions on the degrees, e.g., that all were of the same degree p. He gave elegant formulas, but they don't generalize (perhaps they were too elegant?). After 1910 the subject died out for quite a while, at least in this form. (Giambelli also considered symmetric and skew-symmetric matrices, which are not general in this sense, and found determinantal formulas for these as well. We will discuss this in Chapters 6 and 7.)

Section 1.2 The general problem

Let us consider the general problem of this type. Choose arbitrary upper left submatrices, and put rank conditions on these. We can specify these easily by putting a number r in the (a, b) box of a matrix of boxes, to denote that the upper left $a \times b$ submatrix should have rank at most r.

denotes the locus we looked at in the first example.

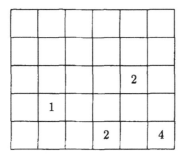

indicates that the upper left 4×2 submatrix has rank at most 1, the upper left 3×5 matrix has rank at most 2, the left 5×4 matrix has rank at most 2, and the whole matrix has rank at most 4.

What can one say about such loci? Consider the simple example

The equations for this locus are $a_{1,1} = 0$, and $a_{2,1} \cdot a_{1,2} = 0$. This means that the locus is the union of the two loci

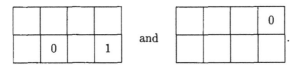

So the locus is the union of two subvarieties of codimension two, to each of which Bézout's theorem applies. Worse, consider the example

This is the union of two loci

The first is irreducible of codimension 5, given by the vanishing of the four left entries and the determinant of the right 2×2 matrix, and the other is irreducible of codimension 4, given by the vanishing of the top row. Again, the degrees of these loci are known by Bézout's theorem.

There is the added problem that one can have many redundant conditions. For example, if there is an r in some box, then putting an r in any box above or to the left of r is clearly redundant, as is putting an $r + 1$ in the box directly below or directly to the right.

We want to describe which loci are irreducible (for generic forms, of positive degree, with N sufficiently large). It is simplest, however, to describe the rank functions of such loci where the rank of every upper left rectangle is specified, whether redundant or not. Let $r(a, b)$ be the rank specified for the upper left $a \times b$ submatrix. Which rank functions r describe irreducible loci? The answer is exactly the same as the answer to another question: exactly which ranks actually *occur* for some $l \times m$ matrix? A simple way to describe the answer is the following. Consider any way to put dots in the boxes, with no two in any row or column. Let $r(a, b)$ be the number of dots in the corresponding upper left $a \times b$ corner. This gives a rank function, and hence, a locus of matrices.

Here is an example with three such dots, the full collection of rank conditions, and the minimal subset of rank conditions:

	•		
		•	
•			

0	1	1	1
0	1	1	1
0	1	1	2
1	2	2	3

			1
0		1	

In general in these notes we will describe some loci in terms of inequalities involving ranks or dimensions. The correct definition, that is necessary for some of the other classical groups, is to define such a locus as the *closure* of the set where these inequalities are equalities. We will also adopt this convention here, since it simplifies the argument.

LEMMA. *Each of the loci corresponding to such a collection of dots is irreducible, and these are exactly the irreducible loci.*

This is proved as follows. The group B^- of lower triangular $l \times l$ matrices acts on the left on the space of all $l \times m$ matrices, preserving any locus described by ranks of upper left corners. And the group B of upper triangular $m \times m$ matrices acts on the right, also preserving any such locus. The matrix obtained by replacing the dots by 1's, and with other entries 0, is a point in the corresponding locus, and left multiplication by B^- and right multiplication by B give all other points in the locus with the same rank equalities. (In down-to-earth terms, one can reduce to one of these standard forms by the elementary row and column operations of multiplying rows or columns by nonzero scalars, and adding a multiple of a row to any row below it, or a multiple of a column to any column to the right.)

This decomposes the space of $l \times m$ matrices into a disjoint union of irreducible locally closed subvarieties. Any locus given by rank conditions must be a union of such loci, and an irreducible such locus must be the closure of one such locus.

The locus described by the rank conditions, in fact, is exactly the closure of the corresponding locally closed set.

Each locus as described in the lemma, or such a collection of dots, corresponds to a unique permutation w in a symmetric group S_n, for some n with $\max(l, m) \leq n \leq l + m$. We may find this w by expanding the array of dots to an $n \times n$ array, adding dots to rows that don't have any, starting from the top down, putting a dot in the left-most column outside the $l \times m$ array that does not yet have a dot. Then set $w(i) = j$ if the dot in the ith row is in the jth column. For the example given before the lemma, the permutation is $w = 25413678$ in S_8, or 25413 in S_5. Here we follow the convention of writing a permutation w by its sequence of values: $w = w(1) w(2) \ldots w(n).$[2] Note that $w(i) < w(i+1)$ if $i > l$ and $w^{-1}(j) < w^{-1}(j+1)$ if $j > m$, and these conditions characterize the permutations that arise from such a collection of dots.

The **length** $l(w)$ of a permutation w is the number of inversions in w, i.e., the number of $i < j$ such that $w(i) > w(j)$.

We denote the locus corresponding to a permutation w by Ω_w. It is the locus where, for each $a \leq l$ and $b \leq m$, the upper left $a \times b$ submatrix of A has rank at most

$$r_w(a, b) = \#\{\, i \leq a : w(i) \leq b \,\}.$$

We will see in the next lecture that Ω_w is an irreducible variety of codimension $l(w)$. The theorem giving its degree didn't appear until 1992, although the ideas for it developed over several decades. We'll give the modern version of this theorem, and sketch the proof, in the next chapter.

THEOREM. *The degree of* Ω_w *is*

$$\mathfrak{S}_w(p_1, p_2, \ldots, p_l, 0, \ldots, 0, -q_1, -q_2, \ldots, -q_m, 0, \ldots, 0),$$

where $\mathfrak{S}_w = \mathfrak{S}_w(x_1, \ldots, x_n, y_1, \ldots, y_n)$ *is the double Schubert polynomial of Lascoux and Schützenberger.*

These **double Schubert polynomials** are characterized (and can be calculated, if not always efficiently) by the following two properties:

(1) If $w = w_0 = n\, n-1 \ldots 2\, 1$, i.e., $w(i) = n + 1 - i$ for $1 \leq i \leq n$, then

$$\mathfrak{S}_w = \mathfrak{S}_w(x_1, \ldots, x_n, y_1, \ldots, y_n) = \prod_{i+j \leq n} (x_i - y_j).$$

(2) If, for some i, $w(i) > w(i + 1)$, and $v = w \cdot s_i$ interchanges the values of w in positions i and $i + 1$, then

$$\mathfrak{S}_v = \partial_i \, \mathfrak{S}_w,$$

---

[2]The convention $w = [w(1), \ldots, w(n)]$ is also common.

where ∂_i is the **divided difference operator** defined on polynomials by the formula

$$(1.8) \qquad\qquad \partial_i P = \frac{P - s_i(P)}{x_i - x_{i+1}},$$

and $s_i(P)$ is obtained from P by interchanging x_i and x_{i+1}. Note that the variables y_j act as scalars in this operation. Since any permutation can be obtained from w_0 by a sequence of such interchanges, (1) and (2) determine all double Schubert polynomials. It is a first fundamental fact about them that they are independent of choices. It is also a fact that one can use any n for which w is in S_n to start this procedure. (As usual, we work in $S_\infty = \bigcup S_n$.)

It is a good idea to work out some of these double Schubert polynomials explicitly, to get an idea of how this works. In order to calculate efficiently, it is useful to note that $\partial_i(P \cdot Q) = P \cdot \partial_i(Q)$ if P is symmetric in the variables x_i and x_{i+1}. More generally we have a formula of Leibniz type:

$$\partial_i(P \cdot Q) = \frac{PQ - s_i(P)s_i(Q)}{x_i - x_{i+1}} = \frac{(P - s_i(P))Q + s_i(P)(Q - s_i(Q))}{x_i - x_{i+1}}$$
$$= \partial_i(P) \cdot Q + s_i(P) \cdot \partial_i(Q).$$

Let's go back to our first example, of 2×3 matrices of rank at most 1. Here one puts one dot in the upper left corner of the 2×3 matrix, to specify that the rank of the whole matrix is at most 1. The corresponding permutation is $w = 1\,4\,2\,3$. One starts with

$$\mathfrak{S}_{4321} = (x_1 - y_1)(x_1 - y_2)(x_1 - y_3)(x_2 - y_1)(x_2 - y_2)(x_3 - y_1).$$

Move the 1 to the left, by interchanging in the third and fourth positions:

$$\mathfrak{S}_{4312} = \partial_3\,\mathfrak{S}_{4321} = (x_1 - y_1)(x_1 - y_2)(x_1 - y_3)(x_2 - y_1)(x_2 - y_2).$$

Then move the 1 another position to the left:

$$\mathfrak{S}_{4132} = \partial_2\,\mathfrak{S}_{4312} = (x_1 - y_1)(x_1 - y_2)(x_1 - y_3)(x_2 + x_3 - y_1 - y_2).$$

Now move the 2 to the left:

$$\mathfrak{S}_{4123} = \partial_3\,\mathfrak{S}_{4132} = (x_1 - y_1)(x_1 - y_2)(x_1 - y_3).$$

Finally, interchange the 1 and the 4:

$$\begin{aligned}
\mathfrak{S}_{1423} &= \partial_1\,\mathfrak{S}_{4123} \\
&= x_1{}^2 + x_1 x_2 + x_2{}^2 - (x_1 + x_2)(y_1 + y_2 + y_3) \\
&\quad + y_1 y_2 + y_1 y_3 + y_2 y_3.
\end{aligned}$$

So we have recovered the formula of Salmon and Roberts.

EXERCISE. Work out the double Schubert polynomials for all permutations in S_4.

Let us consider briefly the problem of finding minimal rank conditions to define a given irreducible locus. In fact, all the rank conditions are determined by their restriction to what we have called in [F2] the **essential set** of the permutation. Any collection of dots, with no two in a row or column, determine a **diagram** D: throw out all boxes lying directly below or to the right of any dot. Note that the rank conditions are constant on connected components of D. It is an elementary lemma that all rank conditions are determined by their restrictions to these components of D, and these are determined by their restrictions to their southeast corners. These boxes form the essential set. In fact, even the ideal generated by the corresponding minors is the same, so that the locus with its full subscheme structure is obtained from the rank functions on these essential sets ([F2, 3.10]). For example, the diagram

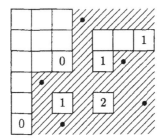

is the same as the diagram $D(w)$ of the permutation w obtained from the dots; in this example, $w = 4\,8\,6\,2\,7\,3\,1\,5$. The length $l(w)$ is the number of boxes in its diagram.

EXERCISE. Show that for $r < l \leq m$, $V_r = \Omega_w$, where

$$w = 1\,2\ldots r\,m+1\,m+2\ldots m+l-r\,r+1\,r+2\ldots m.$$

EXERCISE. Show that a collection of dots corresponds to the rank conditions considered by Giambelli, i.e., rank conditions on full submatrices of sizes $l \times b$ and $a \times m$ for various b and a, if and only if the dots are arranged from northwest to southeast inside the rectangle (equivalently, the corresponding permutation w has $w(i) < w(j)$ whenever $i < j \leq l$ with $w(i) \leq m$ and $w(j) \leq m$).

In Chapter 2 and Appendix B an important class of rank conditions is described, which cover many loci that arise in practise. For these the essential set is easy to find; the corresponding permutations are exactly the permutations called vexillary, and the corresponding Schubert polynomials have particularly nice forms as multi-Schur determinants. This includes in particular the fact that formula (1.4) follows from the theorem.

In general it is far from obvious when an arbitrary collection of squares, together with ranks on them, is such an essential set. K. Eriksson and S. Linusson [E-L] have recently succeeded in giving a complete combinatorial characterization of them.

Section 1.3 More on Schubert polynomials

It is an important property, which follows immediately from the construction, that the double Schubert polynomial $\mathfrak{S}_w(x, y)$ is symmetric in x_i and x_{i+1} if and only if $w(i) < w(i + 1)$. In particular, if $w(i) < w(i + 1)$ for all $i > k$, for some integer k, then the variables x_i for $i > k$ do not appear in $\mathfrak{S}_w$. Although the definition of the operators ∂_i ignores the y variables, these variables occur just as systematically in the double Schubert polynomials. In fact, one has the formula

$$(1.9) \qquad \mathfrak{S}_w(x, y) = (-1)^{l(w)} \mathfrak{S}_{w^{-1}}(y, x).$$

In particular, $\mathfrak{S}_w(x, y)$ is symmetric in y_j and y_{j+1} exactly when $w^{-1}(j)$ is less than $w^{-1}(j + 1)$.

The polynomials $\mathfrak{S}_w(x) = \mathfrak{S}_w(x, 0)$ are called the **Schubert polynomials**, and have been the object of much study by combinatorialists and geometers. They are characterized by the properties that $\mathfrak{S}_{w_0}(x) = x_1{}^{n-1} x_2{}^{n-2} \ldots x_{n-1}$ for $w_0 = n\ n-1\ldots2\ 1$, and $\mathfrak{S}_v(x) = \partial_i \mathfrak{S}_w$ if $v = w \cdot s_i$ and $w(i) > w(i + 1)$. In fact, the plain Schubert polynomials determine the double Schubert polynomials:

$$(1.10) \qquad \mathfrak{S}_w(x, y) = \sum (-1)^{l(v)} \mathfrak{S}_u(x) \cdot \mathfrak{S}_v(y),$$

the sum over all u and v with $v^{-1} \cdot u = w$ and $l(u) + l(v) = l(w)$.

It is a fact that for $w \in S_n$,

$$(1.11) \qquad \mathfrak{S}_w(x) = \sum a_I x^I,$$

the sum over $I = (i_1, \ldots, i_n)$ with $i_j \leq n - j$ for $1 \leq j \leq n$, where x^I denotes $x_1{}^{i_1} \cdot \ldots \cdot x_n{}^{i_n}$, and the coefficients a_I are all *nonnegative* integers. Properties (1.9)–(1.11) and other basic properties of Schubert polynomials can be found in the book of Macdonald [M1].

Billey, Jockusch, and Stanley [B-J-S], cf. [M1, (7.6)], have given an explicit formula for these coefficients:

$$\mathfrak{S}_w(x) = \sum_{a_1, \ldots, a_l} \sum_{i_1, \ldots, i_l} x_{i_1} \cdot \ldots \cdot x_{i_l},$$

where the first sum is over all $a_1, \ldots, a_l$, with $l = l(w)$, such that $w = s_{a_1} \cdot \ldots \cdot s_{a_l}$, and the second sum is over all sequences $i_1, \ldots, i_l$ of positive integers such that $i_1 \leq \cdots \leq i_l$, $i_j \leq a_j$ for all j, and $i_j < i_{j+1}$ if $a_j < a_{j+1}$.

Probably the simplest formula to remember is the following prescription given first by Kohnert. To any diagram D (i.e., a finite collection of boxes numbered by rows and columns as in a matrix), assign the monomial

$$x^D = \prod x_i{}^{\#\text{boxes in the } i^{\text{th}} \text{ row of } D}.$$

A **legal move** on a diagram D chooses the right-most box in any row, and moves it up to the next position above it that has no box in it. The possible legal moves are indicated for a diagram, which is $D(w)$ for $w = 4\,8\,6\,2\,7\,3\,1\,5$.

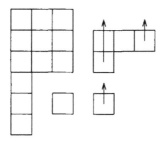

The result of a legal move forms another diagram, on which one can make some new legal moves, and one can repeat until no legal moves are possible. Kohnert's formula is that $\mathfrak{S}_w(x) = \sum x^D$, the sum over all D that arise from $D(w)$ by a sequence of legal moves — each diagram counting once, no matter how many times D can arise this way from $D(w)$. The only problem with this formula is that none of the attempts to prove it have yet been successful!

CHAPTER II

MODERN FORMULATION; GRASSMANNIANS, FLAG VARIETIES, SCHUBERT VARIETIES

Section 2.1 Degeneracy loci of maps of vector bundles

As we have seen, the classical story ended with Giambelli's work in the early part of this century. The modern story started in the 1950's, when R. Thom considered a map $\varphi \colon F \to E$ of complex vector bundles, of ranks m and l, on a variety (then a differentiable manifold, here a complex manifold) X. On open sets where the bundles are trivial, the map is given by a matrix of complex-valued functions on the base. Set

$$D_r(\varphi) = \{\, x \in X : \operatorname{rank}\big(\varphi(x)\big) \le r \,\},$$

a subvariety (usually singular) of X. If the map φ is sufficiently generic, this should have codimension $d = (l-r)(m-r)$, and so should define a cohomology class $[D_r(\varphi)]$ in $H^{2d}(X)$.

The case considered classically, in our first lecture, is that when X is projective space $\mathbb{P}^N$, and $E = \oplus_{i=1}^{l} \mathcal{O}(p_i)$, $F = \oplus_{j=1}^{m} \mathcal{O}(-q_j)$, where $\mathcal{O}(t) = \mathcal{O}(1)^{\otimes t}$, with $\mathcal{O}(1)$ the standard line bundle on $\mathbb{P}^N$. An $l \times m$ matrix A determines a map φ (with A acting on the left on columns), and the locus we called V_r is just $D_r(\varphi)$. Since $H^{2d}(\mathbb{P}^N) = \mathbb{Z} \cdot [L]$, where L is a linear subspace of codimension d, knowing the cohomology class of $D_r(\varphi)$ is the same as knowing its degree.

Thom's basic observation was:

> There must be a polynomial in the Chern classes of E and the Chern classes of F that is equal to this class.

The reason is simple. Consider a "universal" situation. Let G_k be the Grassmannian of rank k subspaces of a large vector space. On $G_m \times G_l$ there are universal bundles E and F of ranks l and m, and a bundle $H = \operatorname{Hom}(F, E) \to G_m \times G_l$, with a universal map φ from F to E on H. (As is common, we omit notation for pullbacks of bundles, here for the maps from H to G_m and H to G_l.) On H, all cohomology classes are polynomials in these Chern classes, so the assertion is clear in this case. The given situation on X is the pullback of the universal map, for some differentiable map from X to H. And the classes and Chern classes pull back as needed, at least if the map is suitably generic.

Thom stated the problem: *Find such a polynomial.* Porteous gave the answer in 1962 (although it was published only in 1971): Define c by the formula

$$c = c(E - F) = c(E)/c(F),$$

where $c(E) = 1 + c_1(E) + c_2(E) + \ldots$ denotes the total Chern class, and the division is carried out formally. Then

$$(2.1) \qquad [D_r(\varphi)] = \Delta_\lambda(c), \qquad \text{where} \qquad \lambda = ((l-r)^{m-r}).$$

Equivalently (by (1.7), or by looking at the dual map $\varphi^\vee \colon E^\vee \to F^\vee$), one can write $[D_r(\varphi)] = \Delta_\mu(c(F^\vee)/c(E^\vee))$, with $\mu = ((m-r)^{l-r})$.

This formula became known as the Thom-Porteous formula. Of course, if mathematicians had been more aware of the classical work on this problem, there would have been no need for the question to be asked — although it would still have required a proof. Now the formula is often called the Giambelli-Thom-Porteous formula.

As in the matrix case, one can give a modern generalization. Suppose we are given a map $h \colon F \to E$ of vector bundles on a nonsingular variety X, and suppose each bundle comes with a partial flag of subbundles. Noting that giving a subbundle is equivalent to giving a quotient bundle, we may write this

$$F_1 \subset F_2 \subset \cdots \subset F_t \subset F \to E \twoheadrightarrow E_s \twoheadrightarrow \cdots \twoheadrightarrow E_2 \twoheadrightarrow E_1.$$

Let w be a permutation such that $w(i) < w(i+1)$ if i is not the rank of any E_a, and $w^{-1}(j) < w^{-1}(j+1)$ if j is not the rank of any F_b. Set

$$\Omega_w(h) = \big\{ x \in X : \mathrm{rank}\big(F_b(x) \to E_a(x)\big) \leq \#\{i \leq \mathrm{rk}(E_a) : w(i) \leq \mathrm{rk}(F_b) \,\forall\, a, b\}.$$

THEOREM 1. *The degeneracy locus $\Omega_w(h)$ is represented by the double Schubert polynomial $\mathfrak{S}_w(x, y)$. In particular, if the map is sufficiently generic, then Ω_w has pure codimension $l(w)$, and $[\Omega_w] = \mathfrak{S}_w(x, y)$.*

Here the x's and y's are the Chern roots of E and F, taken in increasing order by ranks: $x_1, \ldots, x_a$ are the Chern roots for E_1, if $a = rk(E_1)$, $x_{a+1}, \ldots, x_{a'}$ are the Chern roots for $\mathrm{Ker}(E_2 \to E_1)$, etc. And $y_1, \ldots, y_b$ are the Chern roots for F_1, $y_{b+1}, \ldots, y_{b'}$ are the Chern roots for F_2/F_1, etc. The fact that $\mathfrak{S}_w$ is symmetric in the Chern roots of each of these bundles means that it can be expressed in terms of their Chern classes, and it is in this sense that the formula is to be interpreted.

The classical case of matrices of forms is recovered by taking

$$E_a = \mathcal{O}(p_1) \oplus \cdots \oplus \mathcal{O}(p_a) \quad \text{and} \quad F_b = \mathcal{O}(-q_1) \oplus \cdots \oplus \mathcal{O}(-q_b),$$

so the upper left $a \times b$ submatrix of A gives the bundle map from F_b to E_a.

All previously known formulas for degeneracy loci of maps of flagged vector bundles can be deduced from this theorem. This requires some algebra, which says that certain of these double Schubert polynomials can be expressed as certain multi-Schur determinants. The strongest algebraic result we know of this sort is given in Appendix B, and it is this result that we apply next in a geometric setting.

Suppose we have vector bundles

$$B_1 \subset B_2 \subset \cdots \subset B_k \subset B \quad \text{and} \quad A \twoheadrightarrow A_1 \twoheadrightarrow A_2 \twoheadrightarrow \cdots \twoheadrightarrow A_k$$

on a scheme X, of ranks $b_1 \leq b_2 \leq \cdots \leq b_k$ and $a_1 \geq a_2 \geq \cdots \geq a_k$, and we have a morphism $h : B \to A$ of bundles. Let $r_1, \ldots, r_k$ be any sequence of nonnegative integers satisfying

$$0 \leq b_1 - r_1 \leq b_2 - r_2 \leq \cdots \leq b_k - r_k;$$
$$a_1 - r_1 \geq a_2 - r_2 \geq \cdots \geq a_k - r_k \geq 0.$$

Define $\Omega_r(h)$ to be the subscheme of X defined by the conditions that the rank of the map from B_i to A_i is at most r_i, for $1 \leq i \leq k$. Define $p_1, \ldots, p_k$ and $m_1, \ldots, m_k$ by the formulas:

$$p_i = b_{k+1-i} - r_{k+1-i} \qquad \text{for} \quad 1 \leq i \leq k;$$
$$m_1 = a_k - r_k, \quad m_i = (a_{k+1-i} - r_{k+1-i}) - (a_{k+2-i} - r_{k+2-i}) \quad \text{for } 2 \leq i \leq k.$$

Let λ be the partition $(p_1{}^{m_1}, \ldots, p_k{}^{m_k})$. Let $m = m_1 + \cdots + m_k$, and let

$$\Delta_\lambda \big(c(B_k{}^\vee - A_k{}^\vee)^{m_1}, \ldots, c(B_1{}^\vee - A_1{}^\vee)^{m_k} \big)$$

be the determinant of the $m \times m$ matrix $\big(c_{\lambda_i - i + j}(\bullet) \big)$, where, in the first m_1 rows the dot is replaced by $B_k{}^\vee - A_k{}^\vee$, and in the next m_2 rows the dot is replaced by $B_{k-1}{}^\vee - A_{k-1}{}^\vee$, and so on for all the rows of the matrix.

THEOREM 2. *The degeneracy locus $\Omega_r(h)$ is represented by the polynomial*

$$\Delta_\lambda \big(c(B_k{}^\vee - A_k{}^\vee)^{m_1}, \ldots, c(B_1{}^\vee - A_1{}^\vee)^{m_k} \big).$$

There are also dual formulas for these loci. For this, let

$$q_i = a_i - r_i \qquad \text{for} \quad 1 \leq i \leq k;$$
$$n_1 = b_1 - r_1, \quad n_i = (b_i - r_i) - (b_{i-1} - r_{i-1}) \quad \text{for } 2 \leq i \leq k.$$

Let $\mu = (q_1{}^{n_1}, \ldots, q_k{}^{n_k})$.

COROLLARY. *The degeneracy locus $\Omega_r(h)$ is represented by the polynomial*

$$\Delta_\mu \big(c(A_1 - B_1)^{n_1}, \ldots, c(A_k - B_k)^{n_k} \big).$$

We refer to Appendix A for the general meaning of these theorems and corollaries in case the degeneracy locus does not have the expected dimension, or the ambient variety is singular.

The case when each r_i is equal to $b_i - i$ was considered in [P1, (8.3)] and [P4, A.4]. Provided that

$$b_1 - a_1 < b_2 - a_2 < \cdots < b_k - a_k \leq k,$$

the formula for the locus where each $B_i \to A_i$ has kernel of dimension at least i is given by the polynomial

(2.2) $$\Delta_\mu \big(c(A_1 - B_1)^1, \ldots, c(A_k - B_k)^1 \big).$$

Indeed, in this case, $n_i = 1$ for all i, so $\mu = (q_1, \ldots, q_k)$ with $q_i = a_i - b_i + i$, and $q_1 > \cdots > q_k \geq 0$.

Consider the special case of the theorem with all $a_i = a$. In this case one has $B_1 \subset B_2 \subset \cdots \subset B_k \to A$, with ranks $b_1 \leq \cdots \leq b_k$ and a, and one has nonnegative integers r_i satisfying

$$0 \leq b_1 - r_1 \leq b_2 - r_2 \leq \cdots \leq b_k - r_k, \text{ and } r_1 \leq r_2 \leq \cdots \leq r_k \leq a.$$

Then the polynomial for the locus where the rank of $B_i \to A$ is at most r_i is

$$(2.3) \qquad \Delta_\lambda\big(c(B_k{}^\vee - A^\vee)^{m_1}, \ldots, c(B_1{}^\vee - A^\vee)^{m_k}\big)$$
$$= \Delta_\mu\big(c(A - B_1)^{n_1}, \ldots, c(A - B_k)^{n_k}\big),$$

where $\lambda = (p_1{}^{m_1}, \ldots, p_k{}^{m_k})$, with $p_i = b_{k+1-i} - r_{k+1-i}$, $m_1 = a - r_k$, and $m_i = r_{k+2-i} - r_{k+1-i}$ for $2 \leq i \leq k$; $\mu = (q_1{}^{n_1}, \ldots, q_k{}^{n_k})$, where $q_i = a - r_i$, $n_1 = b_1 - r_1$, and $n_i = (b_i - r_i) - (b_{i-1} - r_{i-1})$ for $2 \leq i \leq k$.

In the case where $r_i = b_i - i$ for all i, the dual formula gives exactly the Kempf-Laksov formula ([K-L], cf. [Las3]). This says that if $b_1 < b_2 < \cdots < b_k \leq a + k$, with $\mu_i = a - b_i + i$, then the locus where the dimension of the kernel of $B_i \to A$ is at least i is represented by the polynomial

$$(2.4) \qquad \det\big(c_{a-b_i+j}(A - B_i)\big)_{1 \leq i,j \leq k}.$$

When $k = 1$, i.e., there are no filtrations, one recovers the Giambelli-Thom-Porteous formula: the locus where a map $B \to A$ of vector bundles of ranks b and a has at most rank k, with $k \leq a$ and $k \leq b$, is given by the formula

$$(2.5) \qquad \det\big(c_{b-r-i+j}(B^\vee - A^\vee)\big)_{1 \leq i,j \leq a-r} = \det\big(c_{a-r-i+j}(A - B)\big)_{1 \leq i,j \leq b-r}.$$

Previous proofs of these special cases followed roughly the following pattern, which will be discussed in Chapter 7. Find a variety X' mapping properly to X, on which one has a locus Ω' that maps birationally onto Ω, and for which one can compute its class $[\Omega']$. Usually Ω' is the zero locus of a section of some bundle K of some rank m equal to the codimension of Ω', so $[\Omega'] = c_m(K)$. Then prove a "Gysin formula" to compute the pushforward of $[\Omega']$. We will see that the general theorem is in some ways easier to prove, since one never has to consider anything more complicated than $\mathbb{P}^1$-bundles, where the Gysin formulas are simple.

The loci described in the theorem also include all those considered by Giambelli, where the rank conditions occurred on the bottom row and right column. The permutations that arise in Theorem 2 are what are called **vexillary**: there is no

$$a < b < c < d \qquad \text{with} \qquad w(b) < w(a) < w(d) < w(c).$$

For permutations in S_n, as n grows, vexillary permutations become more and more rare, however. The vexillary permutations are exactly those permutations whose essential sets move across from lower left to upper right, i.e., there are no two of them in positions

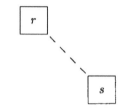

Section 2.2 Grassmannians and flag varieties

There is another important origin of these ideas: the geometry of Grassmannians and flag manifolds. Consider first the Grassmannian $X = G_d V$ of d-dimensional subspaces of a vector space V of dimension n. Fix a complete flag $V_\bullet : V_1 \subset V_2 \subset \cdots \subset V_n = V$, with $\dim(V_i) = i$. For any partition λ whose diagram fits in a d by $n - d$ rectangle, i.e.,

$$n - d \geq \lambda_1 \geq \cdots \geq \lambda_d \geq 0,$$

there is a **Schubert variety**

$$Y_\lambda = Y_\lambda(V_\bullet) = \{L \in X : \dim(L \cap V_{n-d+i-\lambda_i}) \geq i \text{ for } 1 \leq i \leq d\}.$$

Note that when $\lambda_i = 0$, this is an empty condition (which is how one can remember the definition). The claim is that this is an irreducible subvariety of X of codimension $|\lambda| = \sum \lambda_i$, and its class in the cohomology ring of X (which is independent of choice of $V_\bullet$) is given by the "Giambelli formula"

$$(2.6) \qquad\qquad [Y_\lambda] = s_\lambda(x_1, \ldots, x_d) = \Delta_\lambda(c),$$

where $c_i = c_i(E^\vee) = (-1)^i c_i(E)$, with $E \subset V_X$ the **tautological** (or universal) subbundle, and $-x_1, \ldots, -x_d$ are its Chern roots. Here $s_\lambda(x)$ is the Schur polynomial, which can be defined as the corresponding Schur determinant of the complete symmetric functions, cf. (1.6). In fact, it is equal to the Schubert polynomial $\mathfrak{S}_w(x) = \mathfrak{S}_w(x, 0)$, where w is the permutation with $w(i) < w(i+1)$ for $i \neq d$, and with

$$\lambda = (w(d) - d, w(d-1) - (d-1), \ldots, w(2) - 2, w(1) - 1).$$

This is exactly the degeneracy locus Ω_w for the case of

$$(V/V_{n-1})^\vee \hookrightarrow \cdots \hookrightarrow (V/V_2)^\vee \hookrightarrow (V/V_1)^\vee \hookrightarrow V^\vee \twoheadrightarrow E^\vee,$$

which is the dual of

$$E \hookrightarrow V \twoheadrightarrow V/V_1 \twoheadrightarrow V/V_2 \twoheadrightarrow \cdots \twoheadrightarrow V/V_{n-1} \twoheadrightarrow 0.$$

EXERCISE. Verify this, by showing that, for $w(i) = \lambda_{d+1-i} + i$, $L \in \Omega_w \Leftrightarrow$ $\dim(L \cap V_{n-q}) \geq d - \#\{ i \leq d : w(i) \leq q \} \Leftrightarrow L \in Y_\lambda$.

It is a fact that the classes $\sigma_\lambda = [Y_\lambda]$ of these Schubert varieties, called **Schubert classes**, give an additive basis of the cohomology of the Grassmannian. This follows from the fact that each is the closure of an open subvariety — a Schubert cell — that is isomorphic to an affine space, and the union of the Schubert cells in the closure of a Schubert cell is the corresponding Schubert variety. The **Schubert cell** Y_λ° is the locus of subspaces L such that

$$\dim(L \cap V_j) = i \quad \text{for} \quad n - d + i - \lambda_i \leq j \leq n - d + i - \lambda_{i+1},$$

and $\dim(L \cap V_j) = 0$ for $j \leq n - d - \lambda_1$. With V_i spanned by the first i vectors $e_1, \ldots, e_i$ of a basis for V, Y_λ° consists of those subspaces spanned by the rows of an echelon matrix, where the ith row has a 1 in the column numbered $n - d + i - \lambda_i$, and all entries directly below or directly to the right of these 1's are 0. For example, for $n = 10$, $d = 3$, and $\lambda = (3,1,1)$, Y_λ° is described by matrices of the form

$$\begin{bmatrix} * & * & * & * & 1 & 0 & 0 & 0 & 0 & 0 \\ * & * & * & * & 0 & * & * & 1 & 0 & 0 \\ * & * & * & * & 0 & * & * & 0 & 1 & 0 \end{bmatrix},$$

where the $*$'s are arbitrary, and give coordinates identifying Y_λ° with $\mathbb{A}^{16}$. The fact that the classes of the Schubert varieties give a basis for the homology follows from a general fact: whenever a variety has a filtration by closed subspaces such that the successive differences are disjoint unions of spaces each isomorphic to an affine space, then the classes of the closures of these varieties give a basis for the homology (and the Chow) groups of X. It is also a general fact that the class of the variety $Y_\lambda(V_\bullet)$ is independent of choice of fixed flag, since the general linear group acts transitively on the set of such flags, and this continuous group acts trivially on the cohomology. See Appendix A for a brief discussion of these facts.

Classical geometers worked out a good deal of the intersection theory on Grassmannians. For example, Pieri gave a formula for multiplying an arbitrary Schubert class $\sigma_\lambda = [Y_\lambda]$ times a special class $\sigma_{(k)}$. The product $\sigma_{(k)} \cdot \sigma_\lambda$ is the sum of all σ_μ, where μ is obtained from λ by adding k boxes, with no two in the same column. For example, for $n = 10$ and $d = 3$,

$$\sigma_{(3)} \cdot \sigma_{(3,1,1)} = \sigma_{(6,1,1)} + \sigma_{(5,2,1)} + \sigma_{(4,3,1)}.$$

Noting that $\sigma_{(k)} = c_k$, Pieri's formula, together with formula (2.6), gives an algorithm for computing products of Schubert classes.

In the 1930's, in a different context, Littlewood and Richardson gave a closed expression for multiplying two Schur polynomials. Schur polynomials $s_\lambda(x)$ are the characters of basic irreducible representations of $GL_d(\mathbb{C})$ (see Appendix I), and multiplying them is equivalent to decomposing the tensor product of two representations. Their rule gives a combinatorial formula for the coefficient $c_{\lambda \mu}^\nu$ in the formula $s_\lambda \cdot s_\mu = \sum c_{\lambda \mu}^\nu s_\nu$. This rule was only proved many years later, but there are many proofs now (cf. [F6]). Note that the Schur polynomials give an additive basis for all symmetric polynomials, from which it follows that there are such integers $c_{\lambda\mu}^{\nu}$. It follows from representation theory, or the geometry of Grassmannians — since one can move Schubert varieties by the general linear group to get them in general position — that these coefficients are nonnegative. In fact, the Littlewood-Richardson formula shows this nonnegativity, but already for flag manifolds, that we turn to now, no such formula is known.

Classical geometers considered other varieties, such as incidence varieties of pairs of linear subspaces, one contained in another. Ehresmann (1934) and Monk (1959) extended some of this to the variety of complete flags. At about this time the point of view of algebraic groups and homogeneous spaces G/P and G/B began to be studied in earnest, from the point of view of topology (by Borel, Bott, and others), and algebraic geometry (by Chevalley and others — in particular, in an influential paper of Chevalley, written in the 1950's, but only recently published [Ch]).

The complete flag variety $X = Fl(V)$ consists of all flags

$$L_{\bullet} = (L_1 \subset L_2 \subset \cdots \subset L_n = V)$$

with $\dim(L_i) = i$. This is a manifold of dimension $N = \binom{n}{2}$. In fact, it can be constructed by a sequence of projective bundles of ranks $n-1, n-2, \ldots, 2, 1$: start with $\mathbb{P}(V)$, with its tautological line bundle E_1, and then construct $\mathbb{P}(V/E_1) \rightarrow \mathbb{P}(V)$, with its tautological line bundle E_2/E_1, and so on. (As elsewhere in this book, the same notation is often used for a vector space and the corresponding trivial bundle, and for a bundle and its pullback by a given morphism.)

For each w in S_n, and a complete fixed flag $V_{\bullet}$, there is a **Schubert variety** $Y_w = Y_w(V_{\bullet})$ defined by

$$Y_w = \{\, L_{\bullet} \in X : \dim(L_p \cap V_q) \geq \#\{i \leq p : w(i) > n - q\} \ \forall \ p, q \,\}.$$

Its *codimension* is the length of w. There is another common convention for a Schubert variety, that we denote by $X_w = X_w(V_{\bullet})$, defined by the equation

$$X_w = \{\, L_{\bullet} \in X : \dim(L_p \cap V_q) \geq \#\{i \leq p : w(i) \leq q\} \ \forall \ p, q \,\}.$$

This Schubert variety has *dimension* equal to $l(w)$. These are the easiest to visualize. Each X_w is the closure of a Schubert cell X_w°, which is isomorphic to an affine space of dimension $l(w)$. In fact, if one takes a basis for V, and the reference flag has V_k spanned by the first k elements of the basis, then points of X_w° are flags spanned by successive rows of a row echelon matrix, with 1's appearing in the $(i, w(i))$ position, and 0's to the right of and below these 1's. For example, if $w = 5\,2\,6\,1\,3\,4$, then X_w° consists of flags represented by the matrix

$$\begin{bmatrix} * & * & * & * & 1 & 0 \\ * & 1 & 0 & 0 & 0 & 0 \\ * & 0 & * & * & 0 & 1 \\ 1 & 0 & 0 & 0 & 0 & 0 \\ 0 & 0 & 1 & 0 & 0 & 0 \\ 0 & 0 & 0 & 1 & 0 & 0 \end{bmatrix}$$

where the $l(w) = 8$ stars are arbitrary scalars.

Let U_w be the open set in $\mathcal{F}$ spanned by rows of matrices with 1's in the $(i, w(i))$ spots, and 0's below these 1's; taking the other entries as coordinates identifies U_w with $\mathbb{A}^N$, $N = \binom{n}{2}$. In U_w, the equations specified by the rank conditions on the essential set alone exactly make the entries in the diagram $D(w)$ vanish. It follows that $Y_w^\circ = Y_w \cap U_w$ is a linear subspace of U_w of codimension $l(w)$.

The two conventions are related by the equation $Y_w = X_{w_0 \cdot w}$; more precisely, $Y_w(V_\bullet) = X_{w_0 \cdot w}(V_\bullet)$. Again, for the same reason as on the Grassmannian, the classes of the Schubert varieties form an additive basis for the cohomology of X. On X one has a tautological flag of subbundles

$$E_1 \subset E_2 \subset \cdots \subset E_{n-1} \subset E_n = V_X$$

and one has classes $x_i = -c_1(E_i/E_{i-1})$. Since the elementary symmetric polynomials $e_i(x)$ are the Chern classes of the trivial bundle $V_X^\vee$, it follows that $e_i(x) = 0$ in $H^*(X)$. From this it follows that

$$(2.7) \qquad H^*(X) = \mathbb{Z}[x_1, \ldots, x_n]/(e_1(x), \ldots, e_n(x)).$$

One can see this geometrically, using the construction of X as a sequence of projective bundles, or algebraically, using the fact that the ring on the right is a free abelian group with $n!$ generators, e.g., the images of the monomials $x_1^{i_1} \cdot \ldots \cdot x_n^{i_n}$, with $i_j \leq n - j$. These line bundles E_i/E_{i-1} can be given in terms of characters, as in Appendix E, but one must be careful of the sign. An alternative convention is to set $x_i = c_1(E_{n+1-i}/E_{n-i})$.

The problem is to write the classes $[Y_w]$ in terms of these x_i's, and conversely to write polynomials in these x_i's in terms of the Schubert classes. These problems were solved, for general G/B's by Bernstein-Gelfand-Gelfand [B-G-G] and by Demazure [D2]. It was then that the operators ∂_i and their generalizations to other groups first made their appearance. Their answer can be stated in the form

$$(2.8) \qquad [Y_w] = \mathfrak{S}_w(x).$$

This formula, it should be emphasized, takes place entirely in the cohomology ring of the flag manifold, and these authors didn't actually define Schubert polynomials. If one notes that the ring in (2.7) has an additive basis of all monomials of the form $x_1^{i_1} \cdot \ldots \cdot x_n^{i_n}$, where the exponents satisfy the conditions $i_j \leq n - j$, and one writes the representations in terms of this basis, then, in fact, one does find the Schubert polynomials. This amounts to choosing as representative for the class of a point the polynomial

$$\mathfrak{S}_{w_0} = x_1^{n-1} \cdot x_2^{n-2} \cdot \ldots \cdot x_{n-1}.$$

In fact, in [B-G-G] and [D2] other representatives were chosen, which are valid for all simple groups. It was quite a bit later when Lascoux and Schützenberger found and studied these polynomial representatives, for the case when the group is SL_n.

There is an equivalent formulation. For any permutation w, define the operator ∂_w by the formula

$$\partial_w = \partial_{i_1} \circ \cdots \circ \partial_{i_l} \quad \text{if} \quad w = s_{i_1} \cdot \ldots \cdot s_{i_l}, \quad l = l(w).$$

This is independent of choice. Then (2.8) is equivalent to the formula

$$(2.9) \qquad [X_w] = \partial_{w^{-1}}(x_1{}^{n-1} \cdot x_2{}^{n-2} \cdot \ldots \cdot x_{n-1}).$$

Formula (2.8) is a special case of Theorem 1 of Chapter 1, applied to the dual of the situation

$$E_1 \subset E_2 \subset \cdots \subset E_n = V \twoheadrightarrow V/V_1 \twoheadrightarrow V/V_2 \twoheadrightarrow \cdots \twoheadrightarrow V/V_{n-1},$$

where, as is usual, we identify a vector space with the corresponding trivial bundle on any variety. That is, the locus Y_w is the degeneracy locus Ω_w for

$$(V/V_{n-1})^\vee \hookrightarrow \cdots \hookrightarrow (V/V_1)^\vee \hookrightarrow V^\vee \to E_n^\vee \twoheadrightarrow E_{n-1}{}^\vee \twoheadrightarrow \cdots \twoheadrightarrow E_1{}^\vee.$$

EXERCISE. Verify this, by showing that $L_\bullet \in \Omega_w \Leftrightarrow \mathrm{rk}(L_p \to V/V_{n-q}) \le \#\{i \le p : w(i) \le q\} \Leftrightarrow \dim(L_p \cap V_{n-q}) \ge \#\{i \le p : w(i) > q\} \Leftrightarrow L_\bullet \in Y_w$.

The flag manifold X can be used to see more clearly the locus of matrices Ω_w in the space $M_{l,m}$ of matrices satisfying the rank conditions corresponding to a permutation w, i.e., such that the rank of each upper left $p \times q$ submatrix is at most $\#\{i \le p : w(i) \le q\}$. Take $n = l + m$, so w is in S_n. Write $L_\bullet = \langle v_1, \ldots, v_n \rangle$ to denote that L_k is spanned by vectors $v_1, \ldots, v_k$ for $1 \le k \le n$. Let $e_1, \ldots, e_n$ be a basis for V, thereby identifying V with $\mathbb{C}^n$. Let $G = GL_n(\mathbb{C})$, and let $\rho : G \to X$ be the map that takes a matrix A to the flag $\langle v_1, \ldots, v_n \rangle$, where v_i is the ith row of A. This is a smooth, surjective map, that, in fact, makes G into a fiber bundle over X. Let $V_\bullet$ be the flag where V_q is spanned by the last q basis vectors $e_{n+1-q}, \ldots, e_n$. Let $Y_w = Y_w(V_\bullet)$ be the corresponding Schubert variety in X. We have seen that Y_w is a closed irreducible subvariety of X of codimension $l(w)$.

Let $\pi : G \to M_{l,m}$ be the projection that takes a matrix onto its upper left $l \times m$ submatrix. Note that π is also a smooth morphism, surjective since $n \ge l + m$. It follows immediately from the definitions that

$$\pi^{-1}(\Omega_w) = \rho^{-1}(Y_w).$$

From this it follows that Ω_w is also an irreducible subvariety of $M_{l,m}$ of codimension $l(w)$.

Monk's formula is an analogue for flag manifolds of Pieri's formula for Grassmannians. It tells how to multiply a general class $[Y_w]$ by a special class $[Y_{s_i}]$. The result is the sum of all classes $[Y_v]$, where v ranges over all permutations obtained from w by interchanging the values in positions c and d, such that $c \le i < d$, $w(c) < w(d)$, and $w(k)$ is not in $[w(c), w(d)]$ for all k between c and d. For example, for $n = 6$,

$$[Y_{s_3}] \cdot [Y_{241635}] = [Y_{246135}] + [Y_{243615}] + [Y_{261435}]$$
$$+ [Y_{251634}] + [Y_{341625}].$$

In fact, the same formula is valid for multiplying Schubert polynomials, cf. [M1]. See [S1] for generalizations.

In later lectures we will consider the corresponding problems for the other classical groups, i.e., the symplectic and orthogonal groups. We will also consider families, in which one has a vector bundle V, with a nondegenerate bilinear form $V \otimes V \to L$, with values in a line bundle L (often the trivial bundle), that is either symmetric or skew symmetric, and one considers flags of isotropic subbundles. Again there will be degeneracy loci, one for each element of a Weyl group, and our goal is to give formulas for these loci.

Section 2.3 Proof of the theorems

Now let us give a sketch of the proof of Theorem 1. First, it suffices to do the case where the bundles are completely filtered, and the bundle map is the identity:

$$F_1 \subset \cdots \subset F_n = F = E = E_n \twoheadrightarrow E_{n-1} \twoheadrightarrow \cdots \twoheadrightarrow E_1.$$

To see this, given $F \to E$, look at the map $F \subset F \oplus E \twoheadrightarrow E$, where the first map is the graph of the map, and the second is the projection. Then pass to appropriate projective bundles over X to fill in all the steps. (Here one uses the fact that the pullback maps for projective bundles are injective on cohomology or Chow groups.)

Moreover, there is a universal case. Suppose we are given a vector bundle V of rank n and a filtration $F_1 \subset \cdots \subset F_{n-1} \subset V$ on X. Let $\mathcal{F} = Fl(V) \to X$ be the flag bundle, on which one has the tautological flag

$$0 = U_0 \subset U_1 \subset \cdots \subset U_{n-1} \subset U_n = V$$

of subbundles of (the pullback of) V. Let $Q_i = V/U_{n-i}$, so we have the situation

$$F_1 \subset \cdots \subset F_{n-1} \subset V \twoheadrightarrow Q_{n-1} \twoheadrightarrow \cdots \twoheadrightarrow Q_1$$

on $\mathcal{F}$. It suffices to prove the formula in this case, because the previous case is obtained from this by a section $s : X \to \mathcal{F}$. On $\mathcal{F}$ we have, for any w in S_n, the universal subvariety

$$\Omega_w = \big\{ \, x \in \mathcal{F} : \operatorname{rank}(F_q(x) \to Q_p(x)) \leq \#\{i \leq p : w(i) \leq q\} \big\}.$$

This subvariety Ω_w is just a varying version of the Schubert subvariety Y_v of the flag variety that we considered in the preceding section, with $v = w_0 \cdot w \cdot w_0$. The given locus in X is $s^{-1}(\Omega_w)$. In fact, the question of whether a given situation is "generic" enough is transferred to the question of whether this section s is in sufficiently general position with respect to the universal subvariety Ω_w of $\mathcal{F}$; this is discussed in Appendix A.

The formula for the cohomology of a flag manifold generalizes to families. Let $x_i = c_1(\operatorname{Ker}(Q_i \to Q_{i-1})) = c_1(U_{n-i+1}/U_{n-1})$. Then

$$H^*(\mathcal{F}) = H^*(X)[x_1, \ldots, x_n]/\big(e_1(x) - c_1(V), \ldots, e_n(x) - c_n(V)\big).$$

This is easily proved from the construction of $\mathcal{F}$ from X by a sequence of projective bundles, using the standard presentation of the cohomology of a projective bundle over the cohomology of the base. Note that the operators ∂_i determine operators on this ring, with elements in $H^*(X)$ acting as scalars. Let $y_i = c_1(F_i/F_{i-1})$. Our goal is to prove that $[\Omega_w] = \mathfrak{S}_w(x, y)$ for all w. We need two lemmas:

LEMMA 1. $[\Omega_{w_0}] = \prod_{i+j \leq n}(x_i - y_j)$.

This locus is the analogue of a point in the case of the flag manifold. It is the image of the tautological section from X to $\mathcal{F}$, that takes a point x to the flag

$L_\bullet = F_\bullet(x)$. To compute this, one realizes Ω_{w_0} as the zero section of a vector bundle K of rank $n(n-1)/2$ on $\mathcal{F}$:

$$K = \mathrm{Ker}\Big(\bigoplus_{i=1}^{n-1} \mathrm{Hom}(F_i, Q_{n-i}) \to \bigoplus_{i=1}^{n-2} \mathrm{Hom}(F_i, Q_{n-i-1})\Big),$$

where the map takes a collection of homomorphisms from F_i to Q_{n-i} to the differences of the two canonical maps

$$F_i \subset F_{i+1} \to Q_{n-i-1} \qquad \text{and} \qquad F_i \to Q_{n-i} \twoheadrightarrow Q_{n-i-1}.$$

It is not hard to verify that the canonical section of K, given by the maps $F_i \subset V \twoheadrightarrow Q_{n-i}$, vanishes precisely (scheme-theoretically) on the locus Ω_{w_0}, and that its top Chern class is the indicated product.

Next we have to get from the class of this smallest locus to the classes of larger loci. One cannot hope to do this by a *morphism*, but one can look for a *correspondence*. Let $\mathcal{F}(n-i)$ be the partial flag bundle without the $(n-i)^{\text{th}}$ term, i.e., consisting of flags of subspaces of all dimensions except $n-i$. On $\mathcal{F}(n-i)$ there is a universal flag

$$U_1 \subset \cdots \subset U_{n-i-1} \subset U_{n-i+1} \subset \cdots \subset U_n = V,$$

where we have used the same notation as for these bundles as for their pullbacks to $\mathcal{F}$. There is a canonical projection $\mathcal{F} \to \mathcal{F}(n-i)$, which is the $\mathbb{P}^1$-bundle $\mathbb{P}(U_{n-i+1}/U_{n-i-1})$. We have a commutative Cartesian diagram, where all maps are projections of $\mathbb{P}^1$-bundles:

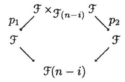

LEMMA 2.
  (1) $p_{1_*} \circ p_2^* : H^{2d}(\mathcal{F}) \to H^{2d+2}(\mathcal{F})$ is equal to ∂_i.
  (2) $p_{1_*} \circ p_2^*[\Omega_w] = [\Omega_{w \cdot s_i}]$ if $w(i) > w(i+1)$.

Both parts of this lemma are straightforward to prove. For the first, one just needs to know about the structure of $\mathbb{P}^1$-bundles. If U is a bundle of rank 2 on a variety Y, $\mathcal{O}(1)$ is the tautological quotient line bundle of U on $\mathbb{P}(U)$, $p : \mathbb{P}(U) \to Y$ is the projection, and $x = c_1(\mathcal{O}(1))$, then any cohomology class in $H^*(\mathbb{P}(U))$ has a unique expression of the form $\alpha x + \beta$, with α and β in $H^*(Y)$. And $p_*(\alpha x + \beta) = \alpha$. Part (1) follows easily from this. (Note that when U corresponds to U_{n-i+1}/U_{n-i-1}, then $\mathcal{O}(1)$ corresponds to U_{n-i+1}/U_{n-i}, so x corresponds to x_i, and $\partial_i(x_i) = 1$.)

Part (2) follows from the fact that, when $w(i) > w(i+1)$, p_1 maps $p_2^{-1}(\Omega_w)$ birationally onto $\Omega_{w \cdot s_i}$. This is a local question, so one is reduced to the case of the flag manifold itself. This is a simple calculation, using the "row echelon" descriptions of Schubert cells, cf. [F2] or [F6]. It is also true that if $w(i) < w(i+1)$ then p_1 maps $p_2^{-1}(\Omega_w)$ into Ω_w, from which it follows that $p_{1_*} \circ p_2^*[\Omega_w] = 0$.

The theorem follows immediately from the two lemmas, together with the two properties (1) and (2) from Chapter 1 that characterize Schubert polynomials.

We next show how Theorem 2 follows from Theorem 1. For this, we need the fact, proved in Appendix B, that the numerical data give rise to a permutation w that is vexillary, and for which the Schubert polynomial can be represented as a multi-Schur polynomial for the partitions λ and μ described there. Take the variables x_i to be the Chern roots of A, and the y_i to be the Chern roots of B, in an order so that the $x_1, \ldots, x_{a_1}$ are the Chern roots for A_1, and $y_1, \ldots, y_{b_1}$ are the Chern roots for B_1, and so on for the successive bundles. Then

$$h_l(a_i, b_i) = c_l(B_i^\vee - A_i^\vee) \qquad \text{and} \qquad e_l(a_i, b_i) = c_l(A_i - B_i).$$

The multi-Schur polynomials appearing in the proposition and its dual in Appendix B are therefore the polynomials in the Chern classes of these bundles required in Theorem 2 and its Corollary.

Since there are different conventions used in the literature, it may be useful here to comment on this. There are two choices of variables x_i and y_i that will give agreement with the established formulas for Schubert polynomials. One is to choose the x_i and y_i as we did in this section (and in [F2]); or one may choose $x_i = -c_1(F_i/F_{i-1})$ and $y_i = -c_1(\mathrm{Ker}(E_i \to E_{i-1}))$ (as was done on the flag manifold in §2.2, and as in [F6]). These two choices are interchanged by the duality that interchanges subbundles and quotient bundles of V and $V^\vee$.

In general, the two involutions $w \mapsto w^{-1}$ and $w \mapsto w_0 \cdot w \cdot w_0$, and the decision about whether to have a Schubert variety labeled by w have dimension $l(w)$ or codimension $l(w)$ had led to several notations for Schubert varieties. Since each of these choices can be useful, it is unlikely that the mathematical world will settle on one convention to the exclusion of all others.

CHAPTER III

SYMMETRIC POLYNOMIALS USEFUL IN GEOMETRY

Most of the topics discussed in this book deal in an essential way with applications of symmetric polynomials to intersection theory. Intersection theory assigns to geometric objects (varieties, subvarieties, schemes, stacks etc.) algebraic objects: groups, rings, ideals, particular elements of rings, in particular polynomials and numbers. For example, we have seen in the previous chapters some formulas for the degree and the fundamental classes which are basic to intersection theory. The most useful tools of intersection theory are characteristic classes associated with vector bundles. Thanks to the splitting principle, these are symmetric polynomials which provide a useful tool to compute with them. On the other hand, symmetric functions constitute a beautiful chapter of both classical and modern algebra. Not all symmetric functions are important for geometry. For instance, the simplest symmetric functions that come to mind, the so-called **monomial symmetric polynomials** m_λ defined for a given partition $\lambda = (\lambda_1, \dots, \lambda_n)$[1] to be the sum of all distinct monomials obtained from $x_1^{\lambda_1} \dots x_n^{\lambda_n}$ by permuting all the variables, have so far played no role in geometry. In this chapter, we will discuss several families of symmetric polynomials whose appearance in geometry is very natural. Usually they appear also in representation theory, although this relationship is not yet fully understood.

Section 3.1 Symmetric polynomials and Grassmannians

Let us start with a brief recollection about the cohomology (or Chow) ring of the Grassmannian $G_d V$ of d-dimensional planes in an n-dimensional complex vector space V. Fix a flag of subspaces

$$V_\bullet : \quad 0 = V_0 \subset V_1 \subset V_2 \subset \dots \subset V_n = V$$

with $\dim(V_i) = i$. Recall that for any partition $\lambda = (\lambda_1, \dots, \lambda_d)$ such that $\lambda_1 \leq n - d$, i.e. $\lambda \subset ((n-d)^d)$, there is a Schubert variety

$$Y_\lambda = Y_\lambda(V_\bullet) = \{L \in G_d V : \dim(L \cap V_{n-d+i-\lambda_i}) \geq i \text{ for } 1 \leq i \leq d\}$$

with its fundamental class $\sigma_\lambda = [Y_\lambda] \in H^{2|\lambda|}(G_d V)$. The following gives a compact description of the intersection ring of $G_d V$ including the classical Schubert Calculus.

---

[1] As is common, we will often use the following simplified notation: $\lambda_1 \lambda_2 \dots \lambda_n$ or $\lambda_1, \dots, \lambda_n$ for a partition $\lambda = (\lambda_1, \dots, \lambda_n)$. This will be especially the case of partitions appearing in lower indices. Moreover, sometimes, it will be convenient to use a notation which indicates the number of times each integer occurs as a part. For example, we denote the partition $(5, 5, 5, 3, 3, 2, 1, 1)$ also by $(5^3, 3^2, 2, 1^2)$ or $5^3 3^2 2 1^2$.

Let us fix a sequence $X = (x_1, \ldots, x_{n-d})$ of $n - d$ variables. Write temporarily $\Delta_\lambda := \Delta_\lambda(e(X))$ with

$$e(X) = 1 + e_1(X) + e_2(X) + \ldots,$$

the sum of elementary symmetric polynomials in X. ($\Delta_\lambda(-)$ is here the Schur determinant from Chapter 1.) The Δ_λ, where λ ranges over all partitions with $\lambda_1 \leq n - d$, form an additive basis of the ring of symmetric polynomials in X. Then the assignment

$$\Delta_\lambda \longmapsto \sigma_\lambda$$

is a *ring* homomorphism and allows us to identify $H^*(G_d V)$ with the quotient ring of the ring $\mathcal{SP}(X)$ of symmetric polynomials in X modulo the ideal $\oplus_\lambda \mathbb{Z}\Delta_\lambda$, where the sum runs over partitions $\lambda \not\subset ((n-d)^d)$. (The fact that this sum is an ideal in the ring $\mathcal{SP}(X)$ follows, e.g., from the Pieri formula.) Moreover, for a "special" Schubert class

$$\sigma_i = [\{L \colon \dim(L \cap V_{n-d+1-i}) \geq 1\}] \in H^{2i}(G_d V),$$

that corresponds here to $\Delta_{(i)}$, we have $\sigma_i = c_i(Q)$ with Q the tautological quotient vector bundle on $G_d V$.

This formulation incorporates all three ingredients of the classical Schubert Calculus mentioned in the previous chapter: the basis theorem, the Pieri formula (which presents the product of a general Schubert class by a special one) and the Giambelli formula (which gives a polynomial presentation of a general Schubert class in terms of special ones): $\sigma_\lambda = \Delta_\lambda(c(Q))$. Also, it tells us how to multiply any two Schubert classes thanks to the Littlewood-Richardson rule. While working over an arbitrary field, the basis theorem remains true for the Chow groups.

There are many variants of this description. As multiplicative generators one can take also the Chern classes $c_i(S)$ of the tautological subbundle S on $G_d V$ or the Segre classes of the tautological bundles; the relations defining $H^*(G_d V)$ can then be expressed through the vanishing of the Chern classes of the remaining tautological vector bundle. For instance, in the above description, since $c_i(S) = 0$ for $i > d$ and, by the Whitney formula $c(Q) \cdot c(S) = 1$, we get the vanishing of the components of degree greater than d of $c(Q)^{-1}$. Consequently, $H^*(G_d V)$ is identified with the ring of symmetric polynomials in X modulo the ideal generated by the complete homogeneous symmetric polynomials $s_i(X)$ where $i > d$.

EXERCISE. Show that $s_i(X)$ for $d < i \leq n$ generate this ideal. Moreover, prove that there exists a symmetric polynomial P of degree $n+1$ such that $(\partial P/\partial e_i)(X) = s_{n+1-i}(X)$ for $i = 1, \ldots, n-d$. (Hint: Consider the power sum polynomial of degree $n+1$.) This presentation of the cohomology ring of the Grassmannian is particularly vivid for a generalization to quantum cohomology. For more on this, see [M-S].

At this point we pause to point out an interesting conjecture concerning endomorphisms of the cohomology rings of the complex Grassmannians, which we believe deserves more attention.

For this discussion, let $X = (x_1, \ldots, x_d)$ be a sequence of d variables. We will use now the identification $H^*(G_d V) \cong \mathcal{SP}(X)/(s_i(X))_{n-d<i\leq n}$, where the elementary symmetric polynomials in X correspond to the Chern classes of the tautological subbundle on $G_d V$.

CONJECTURE. Let f be a degree preserving ring endomorphism of $H^*(G_d V)$. Then $f(e_1(X)) = 0$ implies $f(e_i(X)) = 0$ for all $1 \leq i \leq d$.

Before we discuss this conjecture we mention that non-zero endomorphisms of $H^*(G_d V)$ are fully classified by the following result due to Brewster-Holmer and M. Hoffman (see [Ho] and the references therein). Namely, let f be an endomorphism of $H^*(G_d V)$ with $f(e_1(X)) = me_1(X)$, where $m \neq 0$. Set $e := n - d$. Then, if $d < e$,

$$f(e_i(X)) = m^i e_i(X), \qquad 1 \leq i \leq d.$$

If $d = e$, there is the additional posibility

$$f(e_i(X)) = m^i s_i(X), \qquad 1 \leq i \leq d.$$

(Since a Hermitian inner product on V gives a homeomorphism of $G_d V$ onto $G_e V$ by orthogonal complementation, there is no loss of generality in considering only $G_d V$ with $d \leq e$.)

The proof of this remarkable result is given in [Ho], using the hard Lefschetz theorem. A simpler and more conceptual proof would be welcome.

The conjecture is true if $n \geq 2d^2 - d - 1$ or $d \leq 3$ (see [Gl-Ho]). The conjecture is interesting because $e_1(X)$ does not generate $H^*(G_d V)$ multiplicatively. Although $\{e_1(X), \ldots, e_d(X)\}$ form a minimal set of multiplicative generators of $H^*(G_d V)$, the relations among them seem to force any graded ring endomorphism taking $e_1(X)$ to zero, to send all the $e_i(X)$'s to zero.

The conjecture has the following consequence. If $d < e$ and $d \cdot e$ is even, then every continuous map $\alpha : G_d V \to G_d V$ has a fixed point. Indeed, then

$$\alpha^*(e_i(X)) = m^i e_i(X)$$

for some m, and a rather straightforward computation shows that the Lefschetz number of α is non-zero for any m. Thus α has a fixed point by the Lefschetz fixed-point theorem. For details and more on this subject, we refer the reader to [Gl-Ho] and the references therein.

Just as the Chern classes $c_i(E)$ of a vector bundle E are the elementary symmetric polynomials in the Chern roots of E, there are the so-called Segre classes $s_i(E)$ which are the complete homogeneous symmetric polynomials in these roots.[2] Let us remark that a use of Chern classes is usually more convenient (than Segre classes) in the situation when we look for possibly small, finite set of generators of some algebraic objects: rings, ideals etc. On the other hand, Segre classes are more useful to study Gysin maps in Grassmann or flag bundles, which algebraically correspond to some symmetrizing operators producing complete symmetric polynomials rather than the elementary ones. For this reason we will often switch from Chern to Segre classes and vice versa whenever convenient.

Let us consider now another type of Grassmannians, namely the Lagrangian Grassmannian $LG_n V$ which parametrizes all n-dimensional isotropic subspaces of a $2n$-dimensional complex vector space V endowed with a nondegenerate bilinear

---

[2]The ith Segre class $s_i(E)$ defined here is $(-1)^i$ times that defined in [F1, §4].

skew-symmetric form $\langle \, , \, \rangle$. A subspace L of V is **isotropic** if the form vanishes on it, i.e., $\langle u, v \rangle = 0$ for all u and v in L. A maximal isotropic subspace is called **Lagrangian**.

To describe the additive structure of $H^*(LG_nV)$ introduce a flag

$$V_\bullet : 0 = V_0 \subset V_1 \subset V_2 \subset \ldots \subset V_n \subset V,$$

where V_n is Lagrangian and $\dim(V_i) = i$ for all i.

EXERCISE. Let $e_1, \ldots, e_n$ be a basis of V_n such that $e_1, \ldots, e_i$ is a basis of V_i. Extend this basis to a basis $e_1, \ldots, e_n, f_n, \ldots, f_1$ of V in such a way that $\langle e_i, e_j \rangle = \langle f_i, f_j \rangle = 0$ and $\langle e_i, f_j \rangle = -\langle f_j, e_i \rangle = \delta_{ij}$ for $i, j = 1, \ldots, n$. The Lagrangian Grassmannian LG_nV is a subvariety of the Grassmannian G_nV of all n-dimensional subspaces of V. This last Grassmannian has the decomposition into Schubert cells defined w.r.t. the reference flag whose i^{th} member is spanned by the first i vectors of the above basis. Using row echelon matrices of size $n \times 2n$, analyze, for small values of n ($n = 3, 4$, say), which Schubert cells in G_nV have a nonempty intersection with LG_nV and of which dimension. See also Chapter 6.

Let $\rho(n)$ be the partition $(n, n-1, \ldots, 2, 1)$. A partition λ will be called **strict** if all its parts are distinct. For any strict partition $\lambda = (\lambda_1 > \ldots > \lambda_k > 0) \subset \rho(n)$ (i.e. the parts of λ do not exceed n), there is a **Schubert variety**

$$Y_\lambda = Y_\lambda(V_\bullet) = \{L \in LG_nV : \dim(L \cap V_{n+1-\lambda_i}) \geq i \text{ for } 1 \leq i \leq k\}$$

of codimension $|\lambda|$. As in the case of the ordinary Grassmannian, one Schubert condition defines a **special Schubert variety**

$$Y_i = \{L \in LG_nV : \dim(L \cap V_{n+1-i}) \geq 1\}.$$

Being the closures of "Schubert cells" of LG_nV, the classes of the Y_λ's form an additive basis of $H^*(LG_nV)$. In order to describe the algebraic, and especially multiplicative structure of $H^*(LG_nV)$, we introduce the following "Schur Pfaffian" $\pi_\lambda(c)$.[3] Let $c = c_0 + c_1 + c_2 + \ldots$ be a formal sum of commuting elements where $c_0 = 1$. We set $\pi_i(c) := c_i$, and for $i \geq j \geq 0$ set

$$\pi_{i,j}(c) := \pi_i(c) \cdot \pi_j(c) + 2 \sum_{p=1}^{j} (-1)^p \pi_{i+p}(c) \cdot \pi_{j-p}(c).$$

Given a partition $\lambda = (\lambda_1 \geq \lambda_2 \geq \ldots \geq \lambda_k \geq 0)$, where we can suppose that k is even without loss of generality, we define $\pi_\lambda(c)$ to be the Pfaffian[4] of the $k \times k$ skew-symmetric matrix whose $(p,q)^{th}$ place ($p < q$) is occupied by $\pi_{\lambda_p, \lambda_q}(c)$.

---

[3]To the best of our knowledge, this kind of Pfaffian was first considered by Schur in [Sch].
[4]For the definition and elementary properties of Pfaffians see Appendix D.

EXERCISE.   Show that the so defined $\pi_\lambda(c)$ remains unchanged when we add a string of zeros at the end of λ. (Hint: Use the Laplace-type expansion for Pfaffians (D.1).)

By a Laplace-type expansion for Pfaffians (D.1), the definition of $\pi_\lambda(c)$ can be restated in the form of two recursive relations. Let $\lambda = (\lambda_1 \geq \lambda_2 \geq \ldots \geq \lambda_k > 0)$. If k is odd, then

$$(3.1) \qquad \pi_\lambda(c) = \sum_{p=1}^{k} (-1)^{p-1} \pi_{\lambda_p}(c) \cdot \pi_{\lambda_1,\ldots,\lambda_{p-1},\lambda_{p+1},\ldots,\lambda_k}(c).$$

If k is even, then

$$(3.2) \qquad \pi_\lambda(c) = \sum_{p=2}^{k} (-1)^p \pi_{\lambda_1,\lambda_p}(c) \cdot \pi_{\lambda_2,\ldots,\lambda_{p-1},\lambda_{p+1},\ldots,\lambda_k}(c).$$

For example, $\pi_{43}(c) = c_4 c_3 - 2c_5 c_2 + 2c_6 c_1 - 2c_7$, and $\pi_{421}(c) = c_4 c_2 c_1 - 2c_4 c_3 - 2c_5 c_1 c_1 + 2c_5 c_2 + 2c_6 c_1$.

Given a vector bundle E, we set

$$\widetilde{Q}_\lambda(E) := \pi_\lambda\big(c(E)\big).$$

These polynomials in the Chern classes will play a particularly important role in our story.[5]

If $c = c(1)$ for a formal series $c(T) = 1 + c_1 T + c_2 T^2 + \ldots$ satisfying in addition the equation $c(T) \cdot c(-T) = 1$, then these $\pi_{i,j}(c)$ are the coefficients of the power series in two variables T, U

$$\big(c(T)c(U) - 1\big)\frac{T - U}{T + U} = \sum_{i,j} \pi_{i,j}(c) T^i U^j.$$

Note that $\pi_{i,j}(c) = -\pi_{j,i}(c)$ and this Pfaffian $\pi_\lambda(c)$ vanishes unless λ is a strict partition. We will consider two situations where $c(T) \cdot c(-T) = 1$.

For the first, let $X = (x_1, \ldots, x_n)$ be a sequence of variables and set

$$c(T) := \prod_{i=1}^{n} \frac{1 + x_i T}{1 - x_i T} \quad , \quad c := c(1).$$

In other words, $c = e(X)s(X)$, where $e(X) = 1 + e_1(X) + e_2(X) + \ldots$ is the sum of elementary symmetric polynomials in X and $s(X) = 1 + s_1(X) + s_2(X) + \ldots$ is the sum of complete (homogeneous) polynomials in X. Define

$$Q_\lambda(X) := \pi_\lambda(c).$$

---

[5]Later on, we will also define elements $Q_\lambda(E)$; the notation is justified by the history of the subject.

EXERCISE. Prove that

$$Q_i(X) = 2 \sum_{a+b=i} s_{a1^b}(x_1, \dots, x_n)$$

is twice the sum of all hook Schur polynomials of degree i. (Hint: Use the Pieri formula.)

Here are examples of polynomials $Q_\lambda = Q_\lambda(X)$, for $\lambda \subset \rho(3)$, expressed as combinations of Schur S-polynomials s_λ and monomial symmetric functions m_λ:

$Q_1 = 2s_1 = 2m_1$
$Q_2 = 2s_2 + 2s_{1^2} = 2m_2 + 4m_{1^2}$
$Q_{21} = 4s_{21} = 4m_{21} + 8m_{1^3}$
$Q_3 = 2s_3 + 2s_{21} + 2s_{1^3} = 2m_3 + 4m_{21} + 8m_{1^3}$
$Q_{31} = 4s_{31} + 4s_{2^2} + 4s_{21^2} = 4m_{31} + 8m_{2^2} + 16m_{21^2} + 32m_{1^4}$
$Q_{32} = 4s_{32} + 4s_{31^2} + 4s_{2^21} = 4m_{32} + 8m_{31^2} + 16m_{2^21} + 32m_{21^3} + 64m_{1^5}$
$Q_{321} = 8s_{321} = 8m_{321} + 16m_{31^3} + 16m_{2^3} + 32m_{2^21^2} + 64m_{21^4} + 128m_{1^6}.$

Thus, e.g., we have $Q_{21} = 4m_{21} + 8m_{1^3} = 4\sum x_i x_j^2 + 8\sum_{i<j<k} x_i x_j x_k$.

The polynomials $Q_\lambda(X)$ are called **Schur Q-polynomials** because they were introduced by Schur in the beginning of this century, to describe the projective characters of the symmetric groups. Their algebra has been extensively developed only in recent years (see [H-H]). In particular, we have the following analogue of the Pieri formula for Q-polynomials due to Morris. If $\lambda = (\lambda_1 > \dots > \lambda_k > 0)$ is a strict partition, then

$$Q_r(X) \cdot Q_\lambda(X) = \sum 2^{m(\lambda,\mu)} Q_\mu(X),$$

where the sum is over all strict partitions $\mu = (\mu_1, \dots, \mu_{k+1}) \supset \lambda$ such that $|\mu| = |\lambda| + r$, $\lambda_{p-1} \geq \mu_p \geq \lambda_p$ for $p = 1, \dots, k+1$, and $m(\lambda, \mu)$ is the cardinality of $\{1 \leq p \leq k : \mu_{p+1} < \lambda_p < \mu_p\}$. For example,

$$Q_5 \cdot Q_{6,4,3,2,1} = 2Q_{11,4,3,2,1} + 4Q_{10,5,3,2,1} + 2Q_{9,6,3,2,1} + 4Q_{9,5,4,2,1}$$
$$+ 2Q_{8,6,4,2,1} + 4Q_{8,5,4,3,1} + 2Q_{7,6,4,3,1} + 4Q_{7,5,4,3,2} + Q_{6,5,4,3,2,1}.$$

Observe that, using diagrams, the number $m(\lambda, \mu)$ can be expressed as the number of the connected components of the strip $\mu \smallsetminus \lambda$ not meeting the first column. This is illustrated by the following examples of the summands appearing above with multiplicities 1, 2 and 4 respectively:

$1 \cdot Q_{6,5,4,3,2,1}$ $2 \cdot Q_{7,6,4,3,1}$ $4 \cdot Q_{8,5,4,3,1}$

There exists also a general Littlewood-Richardson-type rule, due to Stembridge, for multiplying two arbitrary Schur Q-polynomials (see e.g. [M2, III.8]). Let us record one consequence of this rule, which will be important for our considerations. Given two strict partitions $\lambda = (\lambda_1 > \ldots > \lambda_k > 0)$ and $\mu = (\mu_1 > \ldots > \mu_l > 0)$, the product $Q_\lambda(X) \cdot Q_\mu(X)$ contains the summand $Q_{\rho(m)}(X)$ with multiplicity 1 if and only if $k + l = m$ and

$$\{\lambda_1, \ldots, \lambda_k, \mu_1, \ldots, \mu_l\} = \{1, 2, \ldots, m\}.$$

(see [P3]). We will write, in such a situation, $\mu = \rho(m) \smallsetminus \lambda$. For example, $(8, 6, 3, 1) = \rho(9) \smallsetminus (9, 7, 5, 4, 2)$.

Consider now the second situation where $c(T) \cdot c(-T) = 1$. Let S be the tautological Lagrangian subbundle and Q the quotient bundle on LG_nV. Then the nondegenerate skew-symmetric form induces an isomorphism $Q \cong S^\vee$, and thus letting $c(T)$ be the Chern polynomial of $S^\vee$, we get $c(T) \cdot c(-T) = 1$. Then, in analogy to the usual Grassmannian case, the Schubert Calculus on LG_nV can be described as follows (this description stems from [P1, (8.7)] and [P3, §6]). For a strict partition $\lambda \subset \rho(n)$, set $\sigma_\lambda := [Y_\lambda] \in H^{2|\lambda|}(LG_nV)$. Let $X = (x_1, \ldots, x_n)$ be a sequence of n variables. Then the assignment

$$Q_\lambda(X) \longmapsto \sigma_\lambda$$

defines a *ring* homomorphism and allows one to identify $H^*(LG_nV)$ with the quotient of the **ring of Schur Q-polynomials** $\mathbb{Z}[Q_1(X), Q_2(X), \ldots] = \oplus_\lambda \mathbb{Z}Q_\lambda(X)$, sum over all strict partitions, modulo the ideal $\oplus_\lambda \mathbb{Z}Q_\lambda(X)$, sum over all partitions $\lambda \not\subset \rho(n)$.

As a consequence of the above mentioned property of Q-polynomials we see that two Schubert classes σ_λ and σ_μ are dual under the pairing $(x, y) \mapsto \int x \cdot y$ of Poincaré duality if and only if $\mu = \rho(n) \smallsetminus \lambda$.

Also, $\sigma_i = c_i(S^\vee)$ and for a general λ we have

$$\sigma_\lambda = \widetilde{Q}_\lambda(S^\vee).$$

This is a Giambelli-type formula for the Lagrangian Grassmannian.

One way to see this identification (and also the one for the usual Grassmannian) is to show that the multiplication tables of the corresponding objects in algebra and geometry are the same. See [P3, §6] for such derivation of the above identification via a comparison of two Pieri-type formulas. Another useful tool for algebraic study of cohomology rings of Grassmannians and flag varieties is provided by the so called *characteristic map*. This group-theoretic approach is explained in Appendix E. It allows one to present the cohomology rings of various homogeneous spaces as some rings of invariants ([B-G-G, §5]). For the Lagrangian Grassmannian this algebraic model for its cohomology ring is $\mathcal{SP}(X)/(e_1(X^2), \ldots, e_n(X^2))$, where $X^2 = (x_1^2, \ldots, x_n^2)$. A Giambelli-type formula for a given Schubert variety Ω becomes in this setup a statement that a certain invariant polynomial of degree equal to codimension of Ω is a unique polynomial annihilated by some family of operators and taking the value 1 under a certain operator canonically associated with Ω.[6] For more on this, we refer the reader to Appendix E, [D1,2], [P-R2-4]

---

[6]Note that such an operator proof of the above identification of the cohomology ring for the Lagrangian Grassmannian with $\widetilde{Q}$-polynomials has appeared only recently in [L-P].

and [P4, §6]. Although if the characteristic map establishes a bridge between rings of invariants and cohomology rings, a more intrinsic explanation of why symmetric polynomials appear in intersection rings would be desirable.

The cohomology (or Chow) rings of the Grassmannians of maximal isotropic subspaces in a $(2n + 1)$-dimensional (resp. $2n$-dimensional) vector space endowed with a nondegenerate bilinear symmetric form admit a similar description (see [P3, §6]).

Finally, we remark that using the σ_i, $i = 1, \ldots, n$, as the multiplicative generators, an alternative description of $H^*(LG_n V)$ is

$$\mathbb{Z}[\sigma_1, \ldots, \sigma_n]/(\sigma_i^2 - 2\sigma_{i+1}\sigma_{i-1} + \ldots + (-1)^i 2\sigma_{2i})_{i=1,\ldots,n},$$

where σ_j is assumed to be zero for $j > n$. This presentation reflects the fact that the only relations among the σ_j's come from the equation $c(S) \cdot c(S^\vee) = 1$. The reader can easily verify that this fact is equivalent to the presentation of $H^*(LG_n V)$ as the quotient ring $\mathcal{SP}(X)/\big(e_1(X^2), \ldots, e_n(X^2)\big)$.

Section 3.2 Symmetric polynomials and degeneracy loci

Let us now pass to degeneracy loci of vector bundle homomorphisms. In the description of their enumerative properties, a prominent role is also played by symmetric polynomials. The most typical is the following situation. Let X be a variety and let $\varphi : F \to E$ be a morphism of vector bundles on X of corresponding ranks m and n. Consider the closed subset of X

$$D_r(\varphi) = \{x \in X : \text{rank}\big(\varphi(x)\big) \leq r\},$$

endowed with a scheme structure defined locally by vanishing of $(r + 1)$-minors of φ. This is rather universal concept overlapping many situations in algebraic geometry (see [F1, §14], [T1]). A natural idea is to try to investigate the enumerative properties of $D_r(\varphi)$ with the help of the Chern classes of the bundles involved. We already know one example of such an enumerative formula: the Giambelli-Thom-Porteous formula. What kind of properties should polynomials in Chern classes of E and F have that are useful for these purposes? Let us start the discussion by mentioning the following problem[7] considered by Harris and Tu in [H-T2]:

> Suppose that X is a compact complex manifold and $D_{r-1}(\varphi) = \emptyset$. Then the kernel K of φ restricted to $D_r(\varphi)$ and its cokernel C are vector bundles of the respective ranks $m - r$ and $n - r$. Find "closed" formulas for the Chern numbers of C and K as polynomials in the Chern classes of E and F. (Note that this problem makes sense when φ is a morphism of C^∞-complex vector bundles.)

Let L be a line bundle on X. Consider the homomorphism $\varphi \oplus 1_L : F \oplus L \to E \oplus L$. Observe that $D_r(\varphi) = D_{r+1}(\varphi \oplus 1_L)$, and that the kernel bundle K and the cokernel bundle C are the same for the two maps φ and $\varphi \oplus 1_L$. Hence the polynomials

---

[7]A solution to this problem will be given in Chapter 5.

evaluating the Chern numbers of K and C should satisfy the following property. Denote by $x_1, \dots, x_n, y_1, \dots, y_m$ and t the Chern roots of E, F and L.

> The polynomial is symmetric in $x_1, \dots, x_n, x_{n+1}$ and $y_1, \dots, y_m, y_{m+1}$ separately, and making the substition $x_{n+1} = y_{m+1} = t$ yields a polynomial not depending on t (thus also obtained by specializing $x_{n+1} = y_{n+1} = 0$).

This observation leads us in a natural way to "Schur polynomials in a difference of alphabets" and/or "supersymmetric polynomials". Let $X = X_n = (x_1, \dots, x_n)$ and $Y = Y_m = (y_1, \dots, y_m)$ be two sequences of independent variables. We say that $F \in \mathbb{Z}[X, Y]$ is **supersymmetric** if F is symmetric in X and Y separately and $F(x_1 = t, y_1 = -t)$ is independent on t (this variant of the independence condition will be used just temporarily to make the life easier by avoiding problems with signs). Here is a family of suppersymmetric polynomials which immediately comes to mind. Define $s_k(X/Y)$ as the coefficients of the series

$$\prod_{i=1}^{n}(1 - x_i T)^{-1} \prod_{j=1}^{m}(1 + y_j T) = \sum s_k(X/Y)T^k.$$

We see that $s_k(X/Y)$ interpolate between $s_k(X)$ (complete homogeneous polynomials in X) and $e_k(Y)$ (elementary symmetric polynomials in Y), and they are supersymmetric. Also supersymmetric is therefore any Schur determinant

$$s_\lambda(X/Y) := \Delta_\lambda(1 + s_1(X/Y) + s_2(X/Y) + \dots) = \det(s_{\lambda_i+j-i}(X/Y)).$$

For example, $s_1(X/Y) = s_1(X) + e_1(Y)$, $s_2(X/Y) = s_2(X) + s_1(X)e_1(Y) + e_2(Y)$, $s_3(X/Y) = s_3(X) + s_2(X)e_1(Y) + s_1(X)e_2(Y) + e_3(Y)$ etc , and $s_{21}(X/Y) = s_2(X/Y)s_1(X/Y) - s_3(X/Y)$.

Observe that, for a partition λ, $s_\lambda(X/0)$ is the classical Schur S-polynomial denoted $s_\lambda(X)$ and originally defined by Jacobi in the form

$$\frac{1}{\Delta(X)} \begin{vmatrix} x_1^{\lambda_1+n-1} & x_1^{\lambda_2+n-2} & \dots & x_1^{\lambda_n} \\ x_2^{\lambda_1+n-1} & x_2^{\lambda_2+n-2} & \dots & x_2^{\lambda_n} \\ \vdots & \vdots & & \vdots \\ x_n^{\lambda_1+n-1} & x_n^{\lambda_2+n-2} & \dots & x_n^{\lambda_n} \end{vmatrix} = \sum_{w \in S_n} w\left[\frac{x_1^{\lambda_1+n-1}x_2^{\lambda_2+n-2}\dots x_n^{\lambda_n}}{\Delta(X)}\right],$$

where $\Delta(X) = \prod_{i<j}(x_i - x_j)$ and S_n acts on $x_1, \dots, x_n$ via permutations. This is perhaps a good moment to introduce some notation which will make many formulas more concise. Given elements $a_1, \dots, a_n$ of a commutative ring and $\alpha = (\alpha_1, \dots, \alpha_n) \in \mathbb{N}^n$, we denote by a^α the monomial $a_1^{\alpha_1} \cdot \dots \cdot a_n^{\alpha_n}$. For two sequences of integers $\alpha = (\alpha_1, \dots, \alpha_n)$ and $\beta = (\beta_1, \dots, \beta_n)$ we will write $\alpha + \beta$ for the sequence $(\alpha_1 + \beta_1, \dots, \alpha_n + \beta_n)$. In particular, using this notation, the above formula becomes

$$s_\lambda(X) = \sum_w w\left[\frac{x^{\lambda+\rho(n-1)}}{\Delta(X)}\right].$$

(Recall that one can always allow one or more zeros to occur at the end of a partition, and identify sequences that differ only by such zeros.)

It turns out that the $\mathbb{Z}$-module of supersymmetric polynomials is freely generated by $\{s_\lambda(X/Y)\}$, where λ runs over the set of partitions whose diagrams D_λ are contained in the (n,m)-hook:

$$D^{n,m} =$$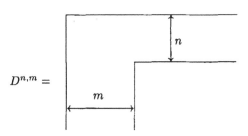

Perhaps the best way to see this, is via the following expression which, we will soon see, is equal to $s_\lambda(X/Y)$. Supposing that $D_\lambda \subset D^{n,m}$, set

$$F_\lambda(X/Y) := \sum_{w \in S_n \times S_m} w \left[\frac{x^{\nu+\rho(n-1)} y^{\mu+\rho(m-1)} \prod\limits_{(i,j)\in D_\lambda^{n,m}} (x_i + y_j)}{\Delta(X)\,\Delta(Y)} \right],$$

where ν, μ and $D_\lambda^{n,m}$ are defined by the following picture displaying D_λ:

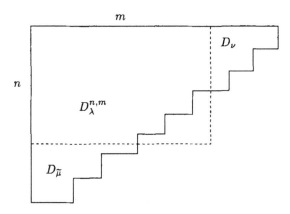

For example, for $\lambda = (4,1,1)$ and $m = n = 2$, we have the picture:

and the above expression becomes

$$\frac{x_1^3 y_1^2 (x_1 + y_1)(x_1 + y_2)(x_2 + y_1)}{(x_1 - x_2)(y_1 - y_2)} + \frac{x_2^3 y_1^2 (x_2 + y_1)(x_2 + y_2)(x_1 + y_1)}{(x_2 - x_1)(y_1 - y_2)}$$

$$+ \frac{x_1^3 y_2^2 (x_1 + y_2)(x_1 + y_1)(x_2 + y_2)}{(x_1 - x_2)(y_2 - y_1)} + \frac{x_2^3 y_2^2 (x_2 + y_2)(x_2 + y_1)(x_1 + y_2)}{(x_2 - x_1)(y_2 - y_1)}.$$

What one can say about this strange expression? Clearly F_λ is a polynomial which is homogeneous of degree $|\lambda|$ and symmetric in X and Y.

EXERCISE. Verify that the polynomial in the above example is supersymmetric.

Here is a general argument. Under the substitution $x_1 = -y_1 = t$, F_λ becomes a polynomial in t and the degree of t in $\Delta(X)\Delta(Y)$ is $n + m - 2$. Hence, if we show that the degree of t in the numerators of all $n!m!$ summands in F_λ cannot exceed $n + m - 2$, then we know that $F_\lambda(x_1 = -y_1 = t)$ doesn't depend on t. Or, equivalently, it suffices to show that for

$$N = x^{\nu + \rho(n-1)} y^{\mu + \rho(m-1)} \prod_{(i,j) \in D_\lambda^{n,m}} (x_i + y_j),$$

the degree of t in $N(x_i = -y_j = t)$ does not exceed $n + m - 2$ for $i = 1, \ldots, n$, $j = 1, \ldots, m$. Of course, if $(i, j) \in D_\lambda^{n,m}$ (matrix coordinates), then after the substitution we get zero. So suppose $(i, j) \notin D_\lambda^{n,m}$. This clearly implies $j > \lambda_i$ and $i > \widetilde{\lambda}_j$ (in particular, $\nu_i = 0$ and $\mu_j = 0$). The factors in N containing x_i or y_j are

$$(x_i + y_1), \ldots, (x_i + y_{\lambda_i}), (x_1 + y_j), \ldots, (x_{\widetilde{\lambda}_j} + y_j), x_i^{n-i}, y_j^{m-j}$$

and the degree of t is therefore

$$\lambda_i + \widetilde{\lambda}_j + (n - i) + (m - j) \le n + m - 2.$$

Note that if $D_\lambda^{n,m}$ is the entire $n \times m$ rectangle, we have the following factorization

$$F_\lambda(X/Y) = s_\nu(X) s_\mu(Y) \prod_{i,j} (x_i + y_j).$$

Using an induction argument, one shows the following result.

PROPOSITION. $\{F_\lambda(X/Y)\}$, where λ runs over the set of partitions with $D_\lambda \subset D^{n,m}$, is a basis of the $\mathbb{Z}$-module of supersymmetric polynomials.

We give a sketch of the proof. First one observes that for the cases $n = 0$ or $m = 0$ starting the induction, the assertion is a simple consequence of the Jacobi presentation of a Schur polynomial and the fact that multiplication by $\Delta(X)$ establishes an isomorphism of the $\mathbb{Z}$-modules of alternating and symmetric polynomials in X. Then one shows, using the above factorization formula, that those F_λ, where $D_\lambda^{n,m}$ is the full $n \times m$-rectangle, form a basis of the $\mathbb{Z}$-module of supersymmetric polynomials which vanish for $x_n = y_m = 0$. To perform the induction step from $(n - 1, m - 1)$ to (n, m), one takes an arbitrary supersymmetric polynomial $P = P(X_n, Y_m)$ and substitutes $x_n = y_m = 0$ in P, thus obtaining a polynomial $P' = P'(X_{n-1}, Y_{m-1})$ which is supersymmetric in X_{n-1} and Y_{m-1}. By noticing that the polynomial $P - P'$ is supersymmetric (in X_n and Y_m) and vanishes under the substitution $x_n = y_m = 0$, one proves the induction assertion by using the induction assumption and the above characterization of supersymmetric polynomials which vanish for $x_n = y_m = 0$. For details, see [P-T].

Finally, we show that

$$F_\lambda(X/Y) = s_\lambda(X/Y)$$

which is, of course, our main goal. To this end, we can assume that $n >> 0$ (in fact, $n \geq |\lambda|$ will do the job). Indeed, letting in either expression some last variables be zero, we get the analogous polynomials associated with λ but depending only on the preceding variables. By the proposition, there exist integers a_σ such that

$$s_\lambda(X/Y) = \sum_\sigma a_\sigma F_\sigma(X/Y),$$

where all partitions σ have n parts at most. Therefore, setting now all the y's to be zero, we do not lose any summand on the right-hand side. But then, invoking the Jacobi presentation of $s_\lambda(X)$ again, we get $a_\lambda = 1$ and $a_\sigma = 0$ for all $\sigma \neq \lambda$.

We have paid so much attention to this formula because it is a key result in the theory of supersymmetric polynomials. It implies immediately

(3.3) $$s_{(m^n)}(X/Y) = \prod_{i,j}(x_i + y_j),$$

and, as we have already noticed, the following factorization formula: for any partition λ with $D_\lambda \subset D^{n,m}$, if $D_\lambda^{m,n}$ is the entire $n \times m$-rectangle,

(3.4) $$s_\lambda(X/Y) = s_\nu(X)s_\mu(Y)s_{(m^n)}(X/Y).$$

EXERCISE. Show that if $D_\lambda \not\subset D^{n,m}$, then $s_\lambda(X_n/Y_m) = 0$. (Hint: Use (3.4).)

Observe that the formula implies also, for any partition λ, the following duality formula

(3.5) $$s_\lambda(X/Y) = s_{\tilde\lambda}(Y/X).$$

Moreover, it "incorporates" one of the central results on Schur S-functions, namely the Littlewood-Richardson rule for multiplying two Schur functions. The interested reader will find a detailed treatment of this application of the formula in [B-G] and [VJ-F]. For more about Schur S-functions, we refer the reader to [M2, I.3] (see also [L-S3]).

Let us now come back to our initial condition of independence under the substitution $x_1 = t = y_1$ imposed by geometry. This is obtained by a change of sign in the basic generating function. Define $s_k(X - Y)$ to be the coefficients of the power series

$$\prod_{i=1}^n(1 - x_iT)^{-1} \prod_{j=1}^m(1 - y_jT) =: \sum_k s_k(X - Y)T^k,$$

and $s_\lambda(X - Y)$ as the Schur determinant

$$s_\lambda(X - Y) := \Delta_\lambda(1 + s_1(X - Y) + s_2(X - Y) + \ldots) = \det\big(s_{\lambda_i+j-i}(X - Y)\big).$$

Of course the family $\{s_\lambda(X-Y)\}$, where λ runs over partitions with $D_\lambda \subset D^{n,m}$, form a basis of the $\mathbb{Z}$-module of polynomials symmetric separately in X and Y and such that the specialization $F(x_1 = t = y_1)$ yields a polynomial independent of t. The polynomials $s_\lambda(X-Y)$ are often called **Schur polynomials in a difference of alphabets**.

Given two vector bundles E and F on a variety X, we define

$$s_\lambda(E-F) := \Delta_\lambda(s(E-F)) = \Delta_\lambda(s(E)/s(F))$$

$$\text{and } \Delta_\lambda(E-F) := \Delta_\lambda(c(E-F)) = \Delta_\lambda(c(E)/c(F)).$$

In other words, letting X, Y be the sequences of Chern roots of E and F, we have $s_\lambda(E-F) = s_\lambda(X-Y)$. We also set

$$s_\lambda(E) := \Delta_\lambda(s(E)).$$

For given two partitions $\lambda = (\lambda_1, \dots, \lambda_l)$ and $\mu = (\mu_1, \dots, \mu_k)$ we denote by λ, μ their juxtaposition, which is the sequence $(\lambda_1, \dots, \lambda_l, \mu_1, \dots, \mu_k)$. Formulas (3.3), (3.4) and (3.5), read, for vector bundles E and F of respective ranks n and m, as follows

$$c_{mn}(E \otimes F^\vee) = s_{(m^n)}(E-F) = \Delta_{(n^m)}(E-F);$$

the **factorization formula**: for a partition $\mu = (\mu_1, \mu_2, \dots)$ such that $\mu_1 \leq m$ and a partition $\nu = (\nu_1, \dots, \nu_n)$,

$$s_{((m^n)+\nu),\mu}(E-F) = s_\nu(E)s_\mu(-F)s_{(m^n)}(E-F)$$
$$= (-1)^{|\mu|} s_\nu(E)s_{\widetilde{\mu}}(F)s_{(m^n)}(E-F);$$

and the **duality formula**: for any partition λ,

$$s_\lambda(E-F) = (-1)^{|\lambda|} s_{\widetilde{\lambda}}(F-E) = \Delta_{\widetilde{\lambda}}(E-F).$$

This set of formulas will be our main algebraic tool in the following two chapters.

As we recall from the previous chapter, the Giambelli-Thom-Porteous formula is evaluated by a polynomial of such a form: given a morphism $\varphi : F \to E$ of vector bundles on a variety X, the degeneracy locus $D_r(\varphi)$ is represented by the polynomial

$$\Delta_{(n-r)^{m-r}}(E-F) = s_{(m-r)^{n-r}}(E-F).$$

Here are some other types of problems that were asked about the enumerative properties of $D_r(\varphi)$:

(1) (Harris-Tu) Assume that X is a compact complex manifold, φ is holomorphic and $D_r(\varphi)$ is smooth. Are there polynomials depending solely on the Chern classes of E, F and X, which evaluate the Chern numbers of $D_r(\varphi)$? Or, without the smoothness assumption of $D_r(\varphi)$ - is there a similar polynomial evaluating the topological Euler-Poincaré characteristic of $D_r(\varphi)$?

(2) What are the homology groups of $D_r(\varphi)$?

(3) Find formulas for the Chern numbers of kernel and cokernel bundles (this problem has already been discussed in some detail in this chapter).

These and some other problems lead in a natural way to considering the following more general problem.

Fix integers $m > 0, n > 0$ and $r \geq 0$. Let $c_\bullet = (c_1, \ldots, c_n)$, $c'_\bullet = (c'_1, \ldots, c'_m)$ be two sequences of independent variables. We say that a polynomial $P \in \mathbb{Z}[c_\bullet, c'_\bullet]$ is **universally supported on the rth degeneracy locus** if for any complex variety X, any homomorphism $\varphi : F \to E$ of vector bundles of ranks m and n on X and any $\alpha \in H_*(X)$,

$$P(c_1(E), \ldots, c_n(E), c_1(F), \ldots, c_m(F)) \cap \alpha \in \mathrm{Im}(\iota_*),$$

where $\iota : D_r(\varphi) \to X$ is the inclusion and $\iota_* : H_*(D_r(\varphi)) \to H_*(X)$ is the induced homomorphism of the homology groups.

It follows directly from the projection formula for ι that polynomials universally supported on the rth degeneracy locus form an ideal, denoted $\mathcal{P}_r$, in $\mathbb{Z}[c_\bullet, c'_\bullet]$. For example, the polynomials

$$\Delta_{(n-k)^{m-k}}\left((1 + c_1 + c_2 + \ldots + c_n)/(1 + c'_1 + c'_2 + \ldots + c'_m)\right)$$

for $0 \leq k \leq r$, are seen to be in $\mathcal{P}_r$. Indeed, by the Giambelli-Thom-Porteous formula, this polynomial is universally supported on the kth degeneracy locus, and thus also on the rth one. These polynomials, however, do not generate the ideals $\mathcal{P}_r$ for $r \geq 1$. Our problem is:

Give an explicit algebraic description of the ideals $\mathcal{P}_r$.

We will deal with this problem in the next chapter. Then we will derive some applications to the enumerative geometry of degeneracy loci.

CHAPTER IV

POLYNOMIALS SUPPORTED ON DEGENERACY LOCI

In order to describe the ideals $\mathcal{P}_r$, we use two tools: the Schur polynomials in the difference of alphabets discussed in the previous chapter and Gysin maps that we will discuss now.

Section 4.1 Gysin maps

Let $E \to X$ be a vector bundle of rank n over a variety X. Let $\pi : G^1E \to X$ be the projective bundle parametrizing rank 1 quotients of E, endowed with the tautological exact sequence

$$0 \longrightarrow S \longrightarrow \pi^*E \longrightarrow \mathcal{O}(1) \longrightarrow 0.$$

Let $a_1 = \xi = c_1(\mathcal{O}(1))$ and $a_2, \dots, a_n$ be the Chern roots of S. It is well known that $\pi_*(\xi^k) = s_{k-(n-1)}(E)$.

EXERCISE. Verify this equality. (See [F1, §4]).

As a matter of fact, this simple formula with the Segre class on the right-hand side is a reason why in this chapter we will work with Segre classes and polynomials in them, rather than with Chern classes.[1]

Let $X_n = (x_1, \dots, x_n)$ be a sequence of variables. Consider the following symmetrizing operator (of Lagrange), defined on $\mathbb{Z}[X_n]^{S_1 \times S_{n-1}}$ by

$$\Delta P := \sum w \left[\frac{P}{(x_1 - x_2)(x_1 - x_3) \dots (x_1 - x_n)} \right],$$

where the sum is taken over $w \in S_n/S_1 \times S_{n-1}$. We claim that π_* is induced by Δ, by which we mean that for any polynomial $P \in \mathbb{Z}[X_n]^{S_1 \times S_{n-1}}$,

$$\pi_*(P(a_1, \dots, a_n)) = (\Delta P)(a_1, \dots, a_n).$$

To see this, one can assume that X is a "sufficiently big" Grassmannian and E is the universal vector bundle on it. Since $\Delta(w(P)) = w(\Delta(P))$ for $w \in S_n$, Δ induces an $H^*(X)$-morphism. Hence, using the decomposition $H^*(G^1E) = \oplus_{i=0}^{n-1} H^*(X)\xi^i$, it sufficient to check that $\Delta(x_1^{n-1}) = 1$.

---

[1] Remember that the present $s_i(E)$ differs from the one in [F1] by a sign.

EXERCISE. Verify this equality. (Hint: It is equivalent to a Laplace expansion of the Vandermonde determinant $\det(x_i^{n-j})_{1 \le i,j \le n}$.)

Let now $\tau = \tau_E : Fl(E) \to X$ be the flag bundle parametrizing successive quotients of ranks $n-1, \ldots, 2, 1$ of E. Let

$$E = Q_n \twoheadrightarrow Q_{n-1} \twoheadrightarrow \ldots \twoheadrightarrow Q_2 \twoheadrightarrow Q_1$$

be the tautological sequence on $Fl(E)$ where $\operatorname{rank}(Q_i) = i$. Set $L_i = \operatorname{Ker}(Q_i \to Q_{i-1})$ and $a_i = c_1(L_i)$ for $i = 1, \ldots, n$. Consider the following **Jacobi symmetrizer**

$$\partial : \mathbb{Z}[X_n] \to \mathbb{Z}[X_n]^{S_n},$$

defined for a polynomial $P \in \mathbb{Z}[X_n]$ by

$$\partial P := \sum w \Big[\frac{P}{\Delta(X_n)} \Big],$$

where we sum over all permutations w in S_n; recall that $\Delta(X_n) = \prod_{i<j}(x_i - x_j)$. This operator induces τ_*, that is

$$\tau_*(P(a_1, \ldots, a_n)) = (\partial P)(a_1, \ldots, a_n).$$

To see this, let us present $\tau = \tau_E : Fl(E) \to X$ as the following composition of two maps

$$Fl(E) = Fl(Q_{n-1}) \xrightarrow{\tau' = \tau_{Q_{n-1}}} G_1 E \xrightarrow{\pi'} X,$$

where $\pi' : G_1 E \to X$ is here the projective bundle parametrizing rank 1 subbundles of E. Consequently, π'_* is induced by the symmetrizing operator Δ', defined for a polynomial $P \in \mathbb{Z}[X_n]^{S_{n-1} \times S_1}$ by

$$\Delta' P = \sum w \Big[\frac{P}{(x_1 - x_n)(x_2 - x_n)\ldots(x_{n-1} - x_n)} \Big],$$

where the summation is over $w \in S_n/S_{n-1} \times S_1$. By induction on n, we can assume that τ'_* is induced by the Jacobi symmetrizer

$$\partial' : P \longmapsto \sum w \Big[\frac{P}{\Delta(X_{n-1})} \Big],$$

the sum over $w \in S_{n-1} \times S_1$. Since $\tau = \pi' \circ \tau'$ and we have $\partial = \Delta' \circ \partial'$, we infer that τ_* is induced by ∂.

The above induction argument can be also applied to establish the following useful identity. If λ is a partition or a sequence $\lambda \in \mathbb{Z}^n$ with $\lambda_i \ge -(n-i)$, for all i,[2] then

$$(4.1) \qquad s_\lambda(E) = \Delta_\lambda(s(E)) = (\partial x^{\lambda + \rho(n-1)})(a_1, \ldots, a_n) = \tau_*(a^{\lambda + \rho(n-1)}).$$

---

[2] If $\lambda_i < -(n-i)$ for some i, it follows from the definition that $\Delta_\lambda(c) = 0$ for any c, so $s_\lambda(E) = 0$.

The middle equality in (4.1), often called the **Jacobi-Trudi identity**, can be shown by induction on n using the above factorization $\tau = \tau' \circ \pi'$ (or equivalently $\partial = \Delta' \circ \partial'$). For details, we refer to [P4, App. 5] (where "Δ" should read "Δ'").

Let us derive from (4.1) a useful formula for Grassmann bundles. Let $\mathcal{G} = G^q E$ be the Grassmannian parametrizing rank q quotients of E, endowed with a tautological sequence

$$0 \longrightarrow S \longrightarrow E_\mathcal{G} \longrightarrow Q \longrightarrow 0.$$

Set $r = n - q$. Then for any partitions $\lambda = (\lambda_1, \ldots, \lambda_q)$, $\mu = (\mu_1, \ldots, \mu_r)$,

$$(4.2) \qquad \pi_*\big(s_\lambda(Q) \cdot s_\mu(S)\big) = s_{\lambda_1-r,\ldots,\lambda_q-r,\mu_1,\ldots,\mu_r}(E).$$

In case the sequence $(\lambda_1-r, \ldots, \lambda_q-r, \mu_1, \ldots, \mu_r)$ is not a partition, the right-hand side is either 0 or $\pm s_\nu(E)$ for some partition ν.[3]

Consider the following commutative diagram

$$
\begin{array}{ccc}
Fl(Q) \times_\mathcal{G} Fl(S) & \xrightarrow{\;\sim\;} & Fl(E) \\
\downarrow{\scriptstyle \tau_Q \times \tau_S} & & \downarrow{\scriptstyle \tau = \tau_E} \\
\mathcal{G} & \xrightarrow{\quad \pi \quad} & X.
\end{array}
$$

Using (4.1) three times, we have

$$
\begin{aligned}
\pi_*\big(s_\lambda(Q) \cdot s_\mu(S)\big) &= \pi_*(\tau_Q \times \tau_S)_*(a_1^{\lambda_1+q-1} \ldots a_q^{\lambda_q} a_{q+1}^{\mu_1+r-1} \ldots a_n^{\mu_r}) \\
&= \tau_*(a_1^{\lambda_1+q-1} \ldots a_q^{\lambda_q} a_{q+1}^{\mu_1+r-1} \ldots a_n^{\mu_r}) \\
&= \tau_*(a_1^{(\lambda_1-r)+(n-1)} \ldots a_q^{(\lambda_q-r)+(n-q)} a_{q+1}^{\mu_1+r-1} \ldots a_n^{\mu_r}) \\
&= s_{\lambda_1-r,\ldots,\lambda_q-r,\mu_1,\ldots,\mu_r}(E).
\end{aligned}
$$

For another proof of the formula (4.2) with the help of a commutative diagram of Grassmann bundles, see [J-L-P]. The method used in [J-L-P] allows us to obtain the following more general result stemming from [P1, §2], whose proof and some generalizations can be found in Appendix F. The setup is the same as in (4.2).

PROPOSITION. *Let F be a vector bundle on X. (i) Let $\lambda = (\lambda_1, \ldots, \lambda_q)$, $\mu = (\mu_1, \ldots, \mu_r)$ be two partitions. Then*

$$(4.3) \qquad \pi_*\big(s_\lambda(Q - F_\mathcal{G}) \cdot s_\mu(S - F_\mathcal{G})\big) = s_{\lambda_1-r,\ldots,\lambda_q-r,\mu_1,\ldots,\mu_r}(E - F).$$

(ii) Let $\lambda = (\lambda_1, \ldots, \lambda_q)$ be a partition such that $\lambda_q \geq \mathrm{rank}(F)$ and $\mu = (\mu_1, \mu_2, \ldots)$ an arbitrary partition. Then

$$(4.4) \qquad \pi_*\big(s_\lambda(Q - F_\mathcal{G}) \cdot s_\mu(S - F_\mathcal{G})\big) = s_{\lambda_1-r,\ldots,\lambda_q-r,\mu}(E - F).$$

---

[3]For a sequence $\lambda = (\lambda_1, \ldots, \lambda_n)$ that is not a partition, a Schur determinant $\Delta_\lambda(c)$ is either 0 or $\pm\Delta_\mu(c)$ for some partition μ. One can rearrange λ by a sequence of operations $(\ldots, i, j, \ldots) \mapsto (\ldots, j-1, i+1, \ldots)$ applied to pairs of successive integers. Either one arrives at a sequence of the form $(\ldots, i, i+1, \ldots)$, in which case $\Delta_\lambda(c) = 0$, or one arrives in d steps at a partition μ, and then $\Delta_\lambda(c) = (-1)^d \Delta_\mu(c)$.

(Recall that given F with sufficiently general Chern classes, $s_\mu(S - F_{\mathcal{G}}) \neq 0$ iff the diagram of μ is contained in the (r, m)-hook $D^{r,m}$, where $m = \text{rank}(F)$.)

We will refer to the formula (4.3) as the **push-forward formula**. One can use this formula to derive push-forward formulas for partial flag bundles. In particular, one obtains the following slight generalization of the Jacobi-Trudi identity. For a partition $\lambda = (\lambda_1, \ldots, \lambda_n)$ one has

$$s_\lambda(E - F) = \tau_*\big(s_{\lambda_1+n-1}(L_1 - F_{\mathcal{F}}) \ldots s_{\lambda_n}(L_n - F_{\mathcal{F}})\big),$$

where $\mathcal{F} = Fl(E)$.

The symmetrizing operator descriptions of the Gysin maps can be generalized to flag bundles associated with principal G-bundles for any semisimple algebraic group G - see Brion's article [Br] and Appendix E. For a divided-difference-operator approach to Gysin maps for flag bundles, consult e.g. [P4, §4]. Algebraic properties of operators which have appeared in this section are developed in [Las2].

Section 4.2 Polynomials universally supported on degeneracy loci

We are now ready to give an explicit description of the ideal $\mathcal{P}_r$, defined at the end of Chapter 3. Let $\mathcal{I}_r \subset \mathbb{Z}[c_\bullet, c'_\bullet]$ be the ideal generated by all polynomials

$$\Delta_\lambda\big((1 + c_1 + c_2 + \ldots + c_n)/(1 + c'_1 + c'_2 + \ldots + c'_m)\big),$$

where $\lambda \supset \big((n - r)^{m-r}\big)$. As a by-product of our method of describing the ideal of polynomials universally supported on the rth degeneracy locus, we will obtain a finite set of generators of $\mathcal{I}_r$.

THEOREM. $\mathcal{P}_r = \mathcal{I}_r$.

We sketch first an algebro-geometric argument proving the inclusion $\mathcal{I}_r \subset \mathcal{P}_r$ asserted by the theorem. It is useful to consider the following geometric construction. Let $G_r E$ be the Grassmann bundle parametrizing r-subbundles of the bundle E, which is equipped with the tautological quotient bundle Q of rank $n - r$. Consider the diagram

$$
\begin{array}{ccc}
Z = Z_r & \xrightarrow{\;j\;} & \mathcal{G} = G_r E \\
\big\downarrow{\eta} & & \big\downarrow{\pi} \\
D_r(\varphi) & \xrightarrow{\;\iota\;} & X
\end{array}
$$

where Z is the scheme of zeros of the composite morphism

$$F_{\mathcal{G}} \xrightarrow{\varphi_{\mathcal{G}}} E_{\mathcal{G}} \xrightarrow{\text{can}} Q,$$

which is embedded by j in $\mathcal{G}$, and η is the restriction of π (note that π restricted to Z factors through $D_r(\varphi)$). Set $q = n - r$. To show that $\mathcal{I}_r \subset \mathcal{P}_r$, take $s_\lambda(E - F)$ where $\lambda \supset \big((m - r)^q\big)$. Then by the push-forward formula,

$$s_\lambda(E - F) \cap \alpha = \pi_*\big(s_{\lambda+(r^q)}(Q - F_{\mathcal{G}}) \cap \pi^*\alpha\big).$$

We have $\lambda + (r^q) \supset (m^q)$. If $\lambda + (r^q) \supset \big((m+1)^{q+1}\big)$, then the diagram of $\lambda + (r^q)$ is not contained in the (q,m)-hook $D^{q,m}$, and thus $s_{\lambda+(r^q)}(Q - F_{\mathcal{G}}) = 0$. In the opposite case, using the factorization formula, we get

$$s_{\lambda+(r^q)}(Q - F_{\mathcal{G}}) = s_{(m^q)}(Q - F_{\mathcal{G}}) \cdot P,$$

where P is a certain polynomial in the Chern classes of Q and $F_{\mathcal{G}}$. Since

$$s_{(m^q)}(Q - F_{\mathcal{G}}) = c_{mq}(F_{\mathcal{G}}^{\vee} \otimes Q)$$

is supported on Z (recall that for any vector bundle A and any section s of it, $c_{\mathrm{top}}(A)$ is supported on the zero scheme of s – cf. Appendix A),

$$s_{(m^q)}(Q - F_{\mathcal{G}}) \cdot P \cap \pi^*(\alpha) = j_*(z)$$

for some $z \in H_*(Z)$. Then

$$s_\lambda(E - F) \cap \alpha = \pi_* j_*(z) = \iota_* \eta_*(z) \in \mathrm{Im}(\iota_*),$$

as claimed.

To show that $\mathcal{P}_r \subset \mathcal{I}_r$, one constructs a sufficiently generic situation where the Chern classes of E and F become algebraically independent. Therefore, if we show that $\mathrm{Im}(\iota_*)$ is equal to the ideal $\mathcal{I}_r$ where $c_i = c_i(E)$ and $c'_j = c_j(F)$, then the required assertion will follow. Let V, W be vector spaces with $v = \dim(V) > n$ and $w = \dim(W) > m$. Set $\mathbb{G} := G^m W \times G_n V$ and denote by Q the tautological quotient bundle on $G^m W$ and by R the tautological subbundle on $G_n V$. Setting

$$H := \mathrm{Hom}(Q_{\mathbb{G}}, R_{\mathbb{G}}) \quad , \quad F' := Q_H \quad , \quad E' := R_H$$

and letting $\varphi' : F' \to E'$ be the tautological homomorphism on H, we get the situation when $c_i(E')$ and $c_j(F')$ become algebraically independent as $v, w \to \infty$. Let X be the Grassmann bundle

$$G_m(Q_{\mathbb{G}} \oplus R_{\mathbb{G}})$$

on $\mathbb{G}$, endowed with the tautological subbundle $S \hookrightarrow (Q \oplus R)_X$. Define $F := S$, $E := R_X$ and φ as the composite map

$$\varphi : F = S \hookrightarrow (Q \oplus R)_X \xrightarrow{\mathrm{can}} R_X = E.$$

Observe that there is an open embedding $H \hookrightarrow X$ of fiber bundles on $\mathbb{G}$, defined over $(M, N) \in \mathbb{G}$ by

$$H_{(M,N)} \ni f \mapsto \big(\mathrm{Graph}(f) \hookrightarrow M \oplus N\big) \in X_{(M,N)}.$$

It is easy to see that φ restricted to H becomes φ', and we get another sufficiently general situation where the Chern classes of E and F become algebraically independent as $v, w \to \infty$. The degeneracy loci $D_k := D_k(\varphi)$ associated with this φ will serve us (also in Chapter 5) as a "generic situation". For this generic situation, one has the following result.

PROPOSITION. *The maps $j^* : H_*(\mathcal{G}) \to H_*(Z)$ and $\eta_* : H_*(Z) \to H_*(D_r)$ are surjective.*

We sketch the proof of this result. Let $Fl_{r,n}V$ be the flag variety parametrizing flags of subspaces of dimension r and n of V. Moreover, let $R^{(r)}$ and $R^{(n)}$ be the tautological bundles of ranks r and n on $Fl_{r,n}V$. To show the first asserted surjection, one uses the fact that the inclusion $j : Z \to \mathcal{G}$ is identified with the following inclusion of Grassmann bundles on $\mathbb{F} = G^m W \times Fl_{r,n}V$:

$$Z = G_m(Q_{\mathbb{F}} \oplus R^{(r)}_{\mathbb{F}}) \hookrightarrow \mathcal{G} = G_m(Q_{\mathbb{F}} \oplus R^{(n)}_{\mathbb{F}}).$$

To prove the second statement, consider $Z^k := \eta^{-1}(D_k)$. Since the D_k's and Z_k's are compact, their singular homology groups are equal to their Borel-Moore homology groups. One shows that the odd Borel-Moore homology groups of D_k and Z^k are zero. Using the fact that $Z^k \smallsetminus Z^{k-1} \to D_k \smallsetminus D_{k-1}$ is a Grassmann bundle, and the homology exact sequences associated with $D_{k-1}, D_k, D_k \smallsetminus D_{k-1}$ and Z^{k-1}, $Z^k, Z^k \smallsetminus Z^{k-1}$ (see Appendix A), one concludes the assertion by induction on k. See [P-R1] for details.

It follows from the second surjection in the proposition that $\text{Im}(\iota_*) = \pi_*(\text{Im } j_*)$. Also, from the first surjection it follows that $\text{Im}(j_*)$ is a principal ideal in $H^*(\mathcal{G})$ generated by $[Z]$. Indeed, let for $z \in H_*(Z)$, $z = j^*(g)$ for $g \in H_*(\mathcal{G})$; then

$$j_*(z) = j_*(j^*g) = [Z] \cdot g.$$

But every element of $H^*(\mathcal{G})$ can be written uniquely in the form

$$g = \sum_\lambda a_\lambda s_\lambda(Q),$$

where $a_\lambda \in H^*(X)$ and $\lambda \subset (r^{n-r})$ (see [F1, §14.6]). Hence we have

$$\pi_*\big([Z] \cdot g\big) = \pi_*\big(s_{(m^{n-r})}(Q - F_{\mathcal{G}}) \cdot \sum_\lambda a_\lambda s_\lambda(Q)\big)$$

$$= \pi_*\big(\sum_\lambda a_\lambda s_{(m^{n-r})+\lambda}(Q - F_{\mathcal{G}})\big)$$

$$= \sum_\lambda a_\lambda s_{(m-r)^{n-r}+\lambda}(E - F)$$

by using successively the factorization formula and the push-forward formula. This last sum belongs to the ideal generated by all $s_{(m-r)^{n-r}+\lambda}(E - F)$ for $\lambda \subset (r^{n-r})$. This ends the sketch of the proof of the inclusion $\mathcal{P}_r \subset \mathcal{I}_r$.[4]

Note that as a by-product of this proof we see that $\mathcal{I}_r$ is generated by the $\binom{n}{r}$ elements

$$\Delta_{(n-r)^{m-r},\lambda}\big((1 + c_1 + \ldots + c_n)/(1 + c'_1 + \ldots + c'_m)\big)$$

where $\lambda \subset ((n-r)^r)$.

CONJECTURE. For $m \geq n$ these elements form a minimal set of generators of the ideal $\mathcal{I}_r$.

---

[4]The theorem was first stated and proved in [P1] for Chow homology. Then it was extended in [P-R1] to Borel-Moore homology and singular homology.

Section 4.3 A generalization of the resultant

It is worth mentioning that the ideal $\mathcal{I}_r$ can be also viewed as a generalization of the *resultant* of two polynomials in one variable. Let

$$f(x) = x^n + \sum_{i=1}^{n}(-1)^i c_i x^{n-i}, \quad g(x) = x^m + \sum_{j=1}^{m}(-1)^j c'_j x^{m-j}$$

be two polynomials in one variable with "generic" coefficients.[5] When do $f(x)$ and $g(x)$, with their coefficients specialized to a field, have a common root in the algebraic closure of that field? An answer given by classical elimination theory (Cayley, Sylvester, ...) says that there is a polynomial in $c_\bullet$ and $c'_\bullet$ called the **resultant** whose vanishing (with the $c_\bullet$ and $c'_\bullet$ specialized in that field) gives a necessary and sufficient condition. Writing in this section

$$\Delta_\lambda = \Delta_\lambda \big((1 + c_1 + \ldots + c_n)/(1 + c'_1 + \ldots + c'_m) \big)$$

for brevity, the resultant can be expressed as $\Delta_{(n^m)}$. One can ask a more general question:

What is the ideal $\mathcal{T}_r \subset \mathbb{Z}[c_\bullet, c'_\bullet]$ of all polynomials P such that for every specialization

$$\alpha : \mathbb{Z}[c_\bullet, c'_\bullet] \to K$$

to a field, if the polynomials

$$x^n + \sum_{i=1}^{n}(-1)^i \alpha(c_i) x^{n-i} \quad \text{and} \quad x^m + \sum_{j=1}^{m}(-1)^j \alpha(c'_j) x^{m-j}$$

have $r + 1$ roots in common in the algebraic closure of that field, then $\alpha(P) = 0$?

Of course, $\mathcal{T}_0$ is a principal ideal generated by the resultant $\Delta_{(n^m)}$.

Surprisingly, classical elimination theory did not give an answer to this question but gave the following two criteria for when the polynomials $f(x)$ and $g(x)$ have $r + 1$ common roots.

The first criterion is due to Trudi (1862): polynomials $f(x)$ and $g(x)$ with coefficients specialized to a field have $r + 1$ common roots in the algebraic closure of that field iff

$$\Delta_{(n-k)^{m-k}} = 0$$

for $k = 0, 1, \ldots, r$.

Pomey (1888) gave another set of polynomials with this property:

$$\Delta_{n-r+k,(n-r)^{m-r-1}}$$

for $k = 0, 1, \ldots, r$.

Both families of polynomials generate $\mathcal{T}_r$ "set-theoretically" (i.e. up to the radical). The whole ideal $\mathcal{T}_r$ is bigger than that generated by these two sets of polynomials. It turns out that for every r, one has:

---

[5]Note that if we write $f(x)$ in this way, then the c_i's become the elementary symmetric polynomials in the roots of $f(x)$.

THEOREM. $\mathcal{T}_r = \mathcal{I}_r$.

The key point in proving this identity is the following result about the structure of $\mathcal{T}_r$.

PROPOSITION. *(1) Every Δ_λ, $\lambda \supset ((n-r)^{m-r})$, is contained in $\mathcal{T}_r$.*
(2) No nonzero $\mathbb{Z}[c_\bullet]$ - combination (resp. $\mathbb{Z}[c_\bullet']$ - combination) of the Δ_λ's with $\lambda \not\supset ((n-r)^{m-r})$, belongs to $\mathcal{T}_r$.

We give an outline of the proof. Let $X = (x_1, \ldots, x_n)$ and $Y = (y_1, \ldots, y_m)$ be two sequences of variables. Then the assignment $c_i \mapsto e_i(X)$, $c_j' \mapsto e_j(Y)$ induces an isomorphism between $\mathbb{Z}[c_\bullet, c_\bullet']$ and the ring $\mathcal{SP}(X|Y) := \mathcal{SP}(X) \otimes_{\mathbb{Z}} \mathcal{SP}(Y)$ of polynomials symmetric separately in X and Y. Thus it is sufficient to prove the two following assertions in $\mathcal{SP}(X|Y)$. (1) Every $s_\lambda(X-Y)$, such that $\lambda \supset ((m-r)^{n-r})$, vanishes under the specialization of X and Y to two sequences having $r+1$ elements in common. Indeed, suppose that $X = (x_1, \ldots, x_{n-(r+1)}, z_1, \ldots, z_{r+1})$ and $Y = (y_1, \ldots, y_{m-(r+1)}, z_1, \ldots, z_{r+1})$. Then, the assertion follows easily from the equality

$$s_\lambda(X-Y) = s_\lambda(X'-Y'),$$

where $X' = (x_1, \ldots, x_{n-(r+1)})$ and $Y' = (y_1, \ldots, y_{m-(r+1)})$. (2) No $\mathcal{SP}(X)$-combination (resp. $\mathcal{SP}(Y)$-combination) of the $s_\lambda(X-Y)$'s with all partitions $\lambda \not\supset ((m-r)^{n-r})$ belongs to $\mathcal{T}_r$. To prove this assertion one uses induction on r. First one shows that the assertion is true if $r = 0$, in which case the ideal $\mathcal{T}_0$ is generated by $s_{(m^n)}(X-Y)$. To show the induction step from $r-1$ to r, one assumes, to the contrary, that the assertion is not true for r. Then there exists a nonzero $\mathcal{SP}(X)$-combination P of the $s_\lambda(X-Y)$'s with all $\lambda \not\supset ((m-r)^{n-r})$, which belongs to $\mathcal{T}_r$. It follows from the induction assumption that P does not belong to $\mathcal{T}_{r-1}$. This implies that P does not vanish under the specialization $X = (x_1, \ldots, x_{n-r}, z_1, \ldots, z_r)$ and $Y = (y_1, \ldots, y_{m-r}, z_1, \ldots, z_r)$ where the x's, y's and z's are three independent over each other sets of variables. But this contradicts the assertion for $r = 0$, for sequences of variables $X' = (x_1, \ldots, x_{n-r})$ and $Y' = (y_1, \ldots, y_{m-r})$ taken instead of X and Y. For details, see [P3, p.164].

It is a simple consequence of the addition theorem[6] for Schur functions that every polynomial from $\mathcal{SP}(X|Y)$ is a combination of the $s_\lambda(X-Y)$'s with coefficients from $\mathcal{SP}(Y)$. To be precise, since $\mathcal{SP}(X|Y) = \mathcal{SP}(X) \otimes_{\mathbb{Z}} \mathcal{SP}(Y)$, it suffices to show the assertion for a polynomial P from $\mathcal{SP}(X)$. We can assume $P = s_\lambda(X)$ and the addition theorem gives $s_\lambda(X) = \sum_\mu s_{\lambda/\mu}(Y) \cdot s_\mu(X-Y)$. Here, $s_{\lambda/\mu}(Y) = \sum_\nu c_{\mu\nu}^\lambda s_\nu(Y)$, where the coefficients $c_{\mu\nu}^\lambda$ come from the formula $s_\mu \cdot s_\nu = \sum_\lambda c_{\mu\nu}^\lambda s_\lambda$ and thus are determined by the Littlewood-Richardson rule. Alternatively, we can require that every polynomial from $\mathcal{SP}(X|Y)$ is a combination of the $s_\lambda(X-Y)$'s with coefficients from $\mathcal{SP}(X)$. Equivalently, every polynomial from $\mathbb{Z}[c_\bullet, c_\bullet']$ is a combination of the Δ_λ's with coefficients from $\mathbb{Z}[c_\bullet']$ (resp. $\mathbb{Z}[c_\bullet]$). Therefore the identity $\mathcal{T}_r = \mathcal{I}_r$ is a consequence of the assertions of the proposition.

---

[6]See [M2, p.74]. Some authors (see e.g. [L-S3]) call this theorem the linearity formula.

Section 4.4 Morphisms with symmetries

There are other types of degeneracy loci that appear in numerous geometric situations. Among those of particular importance are the loci associated with symmetric and skew-symmetric vector bundle morphisms $\varphi : E^\vee \to E$ on a variety X. The definition of symmetric or skew-symmetric can be made by locally trivializing E (and therefore its dual), and looking at the corresponding matrix. Or one can look at the dual map

$$\varphi^\vee : E^\vee \to (E^\vee)^\vee = E;$$

φ is **symmetric** if $\varphi^\vee = \varphi$, and **skew-symmetric** if $\varphi^\vee = -\varphi$. Let

$$D_r(\varphi) = \{x \in X : \operatorname{rank}(\varphi(x)) \leq r\}.$$

In the symmetric case, the scheme structure on $D_r(\varphi)$ is defined by the ideal locally generated by $(r+1)$-minors of φ. In the skew-symmetric case, we assume that r is even, and define the scheme structure by the vanishing of $(r+2)$-subpfaffians of φ. The formulas for the fundamental classes of these degeneracy loci were established in [J-L-P] and [H-T1] and read as follows. In the symmetric case, $D_r(\varphi)$ is represented by the polynomial $2^{n-r}\Delta_{\rho(n-r)}(E)$. In the skew-symmetric case, then $D_r(\varphi)$ is represented by the polynomial $\Delta_{\rho(n-r-1)}(E)$ (recall that r is even).

These formulas will be discussed in Chapter 6 in the context of orthogonal and symplectic degeneracy loci. In this chapter, we are aiming to describe symmetric and skew-symmetric analogs of the ideals $\mathcal{P}_r$. The above polynomials, representing the loci $D_r(\varphi)$, are the generators of the components of minimal degree of these ideals.

The language of Schur Q-polynomials turns out to be an appropriate tool for a description of the corresponding ideals $\mathcal{P}_r^s$, $\mathcal{P}_r^{ss} \subset \mathbb{Z}[c_1, \dots, c_n]$ of polynomials universally supported on r^{th} symmetric (resp. skew-symmetric) degeneracy locus. Their definition is analogous to that for the generic case. For example, we say that a polynomial $P \in \mathbb{Z}[c_\bullet]$ is **universally supported on the rth symmetric degeneracy locus** if for any (complex) variety X, any symmetric homomorphism $\varphi : E^\vee \to E$ on X with $\operatorname{rank}(E) = n$ and any $\alpha \in H_*(X)$,

$$P(c_1(E), \dots, c_n(E)) \cap \alpha \in \operatorname{Im}(\iota_*),$$

where $\iota : D_r(\varphi) \to X$ is the inclusion and $\iota_* : H_*(D_r(\varphi)) \to H_*(X)$ is the induced homomorphism of homology groups.

Given a vector bundle E and a strict partition λ, let

$$Q_\lambda(E) := \pi_\lambda(c(E) \cdot s(E)),$$

where π_λ is the Schur Pfaffian from Chapter 3. In other words, $Q_\lambda(E)$ is $Q_\lambda(X)$ (defined in Chapter 3) after specializing X to the sequence of the Chern roots of E. Set also $P_\lambda(E) := 2^{-l(\lambda)}Q_\lambda(E)$ and $P_\lambda(X) := 2^{-l(\lambda)}Q_\lambda(X)$, where $l(\lambda)$ denotes the **length** of the partition λ, that is the number of its parts. The polynomials $P_\lambda(X)$ and $P_\lambda(E)$ are called **Schur P-polynomials**. Notice that $P_\lambda(X)$ is a polynomial with integer coefficients. For instance, $P_i(X) = \sum s_\lambda(X)$, the sum over all hook partitions λ with $|\lambda| = i$.

Schur Q- and P-polynomials are also useful to establish the enumerative properties of these loci. A reason for that can be explained by the problem of computing the Chern numbers of kernel and cokernel bundles. So let $\varphi : E^{\vee} \to E$ be a symmetric morphism such that $D_{r-1}(\varphi) = \emptyset$. Then, restricted to $D_r(\varphi)$, the morphism φ has the constant rank r and its kernel K and cokernel C are vector bundles of rank $n - r$.

EXERCISE. Show that $C \cong K^{\vee}$.

Let L be a line bundle and set $F := L \oplus L^{\vee}$. Consider the symmetric morphism

$$\varphi \oplus \alpha : E^{\vee} \oplus F^{\vee} \to E \oplus F,$$

where $\alpha : F^{\vee} \to F$ is the symmetric isomorphism defined by the 2×2 matrix

$$\begin{pmatrix} 0 & 1_L \\ 1_L & 0 \end{pmatrix}.$$

Observe that $D_r(\varphi) = D_{r+2}(\varphi \oplus \alpha)$, and that the kernel bundle K is the same for the two morphisms φ and $\varphi \oplus \alpha$ restricted to this degeneracy locus. Therefore the polynomials in the Chern roots of E evaluating the Chern numbers of K should have the property expressed by the following problem:

What are the polynomials symmetric in $x_1, \ldots, x_n, x_{n+1}, x_{n+2}$ such that the specialization $x_{n+1} = t$, $x_{n+2} = -t$ yields a polynomial not depending on t?

It turns out that any such a polynomial is a $\mathbb{Z}$-combination of the $P_\lambda(X)$'s. It is shown in [P3, §2] that this independence property for $P_\lambda(X)$ can be deduced from the following "factorization property" due to Stanley: for any partition λ with $l(\lambda) \leq n$,

$$(4.5) \qquad\qquad P_{\rho(n-1)+\lambda}(X) = P_{\rho(n-1)}(X) \, s_\lambda(X)$$

where, more explicitly,

$$(4.6) \qquad\qquad P_{\rho(n-1)}(X) = \prod_{1 \leq i < j \leq n} (x_i + x_j).$$

These two formulas will be simple consequences of the symmetrizing operator expression for $P_\lambda(X)$ given at the end of this chapter.

Another useful formula is the following formula for the Gysin map. Let $\mathcal{G} = G^q E$ be the Grassmann bundle parametrizing rank q-quotients of E. Then for strict partitions λ and μ with $k = l(\lambda) \leq q$ and $h = l(\mu) \leq r = n - q$,

$$(4.7) \qquad \pi_* \big(c_{qr}(Q \otimes S) \cdot P_\lambda(Q) \cdot P_\mu(S) \big) = d \cdot P_{\lambda_1, \ldots, \lambda_k, \mu_1, \ldots, \mu_h}(E)$$

where $d = 0$ if $(q - k)(r - h)$ is odd and

$$d := (-1)^{(q-k)r} \binom{[(n - k - h)/2]}{[(q - k)/2]}$$

otherwise. It can happen that $(\lambda_1, \ldots, \lambda_k, \mu_1, \ldots, \mu_h)$ is not a strict partition. If there are repetitions of indices in $(\lambda_1, \ldots, \lambda_k, \mu_1, \ldots, \mu_h)$, then the right-hand side is assumed to be zero; if not then

$$P_{\lambda_1, \ldots, \lambda_k, \mu_1, \ldots, \mu_h}(E) = (-1)^l P_\nu(E),$$

where l is the length of the permutation which rearranges $(\lambda_1, \ldots, \lambda_k, \mu_1, \ldots, \mu_h)$ into the corresponding strict partition ν. For a proof of this last push-forward formula, see [P4, App.1].

EXAMPLE. 1) For $n = 15$, $q = 7$, $r = 8$, $k = 3$, $h = 4$, one has

$$\pi_*\big(c_{56}(Q \otimes S) \cdot P_{9,3,1}(Q) \cdot P_{7,5,4,2}(S)\big) = (-6) \cdot P_{9,7,5,4,3,2,1}(E).$$

2) If $l(\lambda) = q - 1$ and $\mathrm{rank}(S)$ is odd, then

$$\pi_*\big(c_{qr}(Q \otimes S) \cdot P_\lambda(Q)\big) = 0.$$

In fact, to obtain an explicit description of $\mathcal{P}_r^s$ and $\mathcal{P}_r^{ss}$ only the following special case is needed:

$$\pi_*\big(c_{qr}(Q \otimes S) \cdot P_\lambda(Q)\big) = P_\lambda(E)$$

if $l(\lambda) = q$ or $l(\lambda) = q - 1$ and $\mathrm{rank}(S)$ is even. Then, following the strategy of the generic case, one arrives at the following description of $\mathcal{P}_r^s$ and $\mathcal{P}_r^{ss}$. Let $c = 1 + c_1 + c_2 + \ldots$, where $c_1, c_2, \ldots$ are variables. Set

$$Q_\lambda := \pi_\lambda(c \cdot s)$$

where $s = 1 + s_1 + s_2 + \ldots$ with s_i defined inductively by

$$s_i - s_{i-1}c_1 + s_{i-2}c_2 - \ldots + (-1)^i c_i = 0.$$

Also, set $P_\lambda := 2^{-l(\lambda)}Q_\lambda$. Then

(4.8)
$$\mathcal{P}_r^s = \big(Q_{\rho(n-r)+\lambda} : \lambda \subset (r^{n-r})\big)$$

and

(4.9)
$$\mathcal{P}_r^{ss} = \big(P_{\rho(n-r-1)+\lambda} : \lambda \subset (r^{n-r})\big)$$

(see [P1, §7] and [P-R1]). In particular, looking at the components of minimal degree of these ideals, we get the following alternative expressions for the polynomials representing the degeneracy loci associated with symmetric and skew-symmetric morphisms. Given a symmetric morphism $\varphi : E^\vee \to E$ on X, the degeneracy locus $D_r(\varphi)$ is represented by the polynomial $Q_{\rho(n-r)}(E)$. Similarly, given a skew-symmetric morphism $\varphi : E^\vee \to E$ on X, the degeneracy locus $D_r(\varphi)$, for even r, is represented by the polynomial $P_{\rho(n-r-1)}(E)$. The effect of comparison of these expressions with those recalled at the beginning of this section corresponds to the following identity in the theory of P-polynomials:

(4.10)
$$P_{\rho(k)}(X) = s_{\rho(k)}(X)$$

for every k.

We refer to [P3, §5] for a discussion of "resultant-like" analogs of $\mathcal{P}_r^s$ and $\mathcal{P}_r^{ss}$.

We end this chapter with two facts related to this story. The first will be used in the next chapter.

PROPOSITION 1. *Consider the Grassmann bundles*

$$\pi_E : G_r E \to X \qquad and \qquad \pi_F : G^r F \to X,$$

where $\mathrm{rank}(E) = n$ *and* $\mathrm{rank}(F) = m$, *equipped with the tautological sequences*

$$0 \to S_E \to \pi_E^* E \to Q_E \to 0 \quad and \quad 0 \to S_F \to \pi_F^* F \to Q_F \to 0.$$

Denote by τ *the composite morphism*

$$\tau : \mathbb{G} = G^r F \times_X G_r E \xrightarrow{\pi_F \times 1} G_r E \xrightarrow{\pi_E} X.$$

Then, for the vector bundle $H = \mathrm{Hom}(F, E)/\mathrm{Hom}(Q_F, S_E)$ *on* $\mathbb{G}$ *(we omit writing the pull-back indices) and two partitions* λ *and* μ *where* $l(\lambda) \leq n-r$ *and* $l(\widetilde{\mu}) \leq m-r$, *one has*

$$(4.11) \qquad \tau_* \Big(s_\lambda(Q_E)\, s_\mu(-S_F)\, c_{\mathrm{rank}(H)}(H) \Big) = s_{((m-r)^{n-r}+\lambda),\mu}(E - F).$$

We give a sketch of the proof. First note that in $K(\mathbb{G})$,

$$[\mathrm{Hom}(F, E)/\mathrm{Hom}(Q_F, S_E)] = [S_F^\vee \otimes S_E] + [F^\vee \otimes Q_E],$$

so we must compute:

$$(4.12) \qquad \tau_* \Big(s_\lambda(Q_E)\, s_\mu(-S_F)\, s_{(m-r)^r}(S_E - S_F)\, s_{(m^{n-r})}(Q_E - F) \Big).$$

Next, applying the duality formula, the factorization formula to the two middle factors in (4.12), and the push-forward formula for π_F, one gets the identity

$$(\pi_F)_* \big(s_\mu(-S_F)\, s_{(m-r)^r}(S_E - S_F) \big) = s_\mu(S_E - F).$$

Using this last identity and the factorization formula to the first and fourth factors in (4.12), one transforms the expression (4.12) to the following one:

$$(\pi_E)_* \big(s_{(m^{n-r})+\lambda}(Q_E - F)\, s_\mu(S_E - F) \big),$$

and one finishes the computation by applying the formula (4.4). (This result stems from [P1, §3].)

The last proposition in this section stems essentially from [Sch].

PROPOSITION 2. *Let* $X = (x_1, \ldots, x_n)$ *be a sequence of variables, and* $\lambda = (\lambda_1, \ldots, \lambda_l)$ *a strict partition of length* $l \leq n$. *Then one has*

$$(4.13) \qquad Q_\lambda(X) = 2^l \sum_{w \in S_n/S_{n-l}} w\Big[x^\lambda \prod \frac{x_i + x_j}{x_i - x_j} \Big],$$

where S_{n-l} acts on $x_{l+1}, \ldots, x_n$, and the product is over pairs (i,j) such that $1 \le i \le l$ and $1 \le i < j \le n$.

For example,

$$Q_r(X) = 2 \sum_{i=1}^{n} x_i^r \prod_{i' \neq i} \frac{x_i + x_{i'}}{x_i - x_{i'}},$$

$$Q_{r,s}(X) = 4 \sum_{i=1,j=1}^{n} x_i^r x_j^s \frac{x_r - x_s}{x_r + x_s} \prod_{i' \neq i} \frac{x_i + x_{i'}}{x_i - x_{i'}} \prod_{j' \neq j} \frac{x_j + x_{j'}}{x_j - x_{j'}}.$$

We give an outline of the proof. For $l = 1$, one deduces the expression directly from the generating series defining the Q_r's by using the usual rule for partial fractions. For $l = 2$, one verifies it by induction on s using the relation

$$Q_{r,s}(X) = Q_r(X)Q_s(X) - Q_{r+1}(X)Q_{s-1}(X) - Q_{r+1,s-1}(X).$$

For $l \ge 3$, one shows that the above expressions (4.13) satisfy the recursion relations (3.1) and (3.2), defining the $Q_\lambda(X)$'s. For more details, see [H-H] or [Sch].

One deduces easily from (4.13) the factorization property (4.5) and the identity (4.6)

$$P_{\rho(n-1)}(X_n) = \prod_{i<j} (x_i + x_j).$$

In fact, the push-forward formula (4.7) is also a consequence of this symmetrization operator formula (see [P4, App.1]).

CHAPTER V

THE EULER CHARACTERISTIC OF DEGENERACY LOCI

In this chapter we will apply the technique of polynomials universally supported on degeneracy loci to compute the topological Euler-Poincaré characteristic (Euler characteristic for short) of degeneracy loci. The final formula will express the Euler characteristic of $D_r(\varphi)$, for a (holomorphic) homomorphism $\varphi : F \to E$ of vector bundles on a compact complex manifold X, as a polynomial depending solely on the Chern classes of E, F and TX. It will hold, however, only under strong transversality assumptions on the morphism φ.

Section 5.1 Sections of bundles and general bundle morphisms

Let X, Y be complex manifolds and Z a submanifold of Y. Recall that a differentiable map $f : X \to Y$ is *transverse* to Z if at any point $x \in X$ with $f(x) \in Z$, the composition

$$T_x X \xrightarrow{(df)_x} T_{f(x)} Y \xrightarrow{\text{can}} N_{f(x)}$$

is surjective. Here, $(df)_x$ is the differential of f at x and $N_{f(x)}$ is the normal space to Z in Y at $f(x)$. Then $f^{-1}Z$ is nonsingular and $\text{codim}(f^{-1}Z, X) = \text{codim}(Z, Y)$. For more on this, see [G-G]. For us, in this section, the most important will be the concept of a section of a vector bundle, meeting transversally (the image of) the zero section of the bundle.

EXERCISE. Let E be a vector bundle on a complex manifold X and $s : X \to E$ its section. Suppose that s is transverse to (the image of) the zero section of E. Denoting by Z the subvariety of zeros of s, show that the normal bundle $N_{Z/X}$ of Z in X is canonically isomorphic to $E|_Z$.

In algebraic terms, the transversality of s to the zero section of a bundle E of rank n is expressed as follows: if s is given in local coordinates around a point $x \in X$ by $(s_1, \ldots, s_n)$, $s_i \in \mathcal{O}_{x,X}$, then $s_1, \ldots, s_n$ form a part of a regular system of parameters of $\mathcal{O}_{x,X}$. In particular, if the subvariety of zeros of s is nonsingular of codimension n, then s is transverse to the zero section.

We want to apply the classical Gauss-Bonnet formula (or Hopf's theorem) asserting that for a compact complex manifold X, the Euler characteristic $\chi(X)$ of X is evaluated as:

$$\chi(X) = \int_X c(X),$$

where $c(X)$ is the total Chern class of the tangent bundle TX to X. (The right-hand side of this formula means the degree of the top Chern class of TX.)

Assume that X is a compact complex manifold and we are in the situation of the exercise. Since

$$[TZ] = [TX|_Z] - [E|_Z]$$

in $K(Z)$, the Gauss-Bonnet formula implies that

$$\chi(Z) = \int_X c_{\text{top}}(E)\, c(E)^{-1}\, c(X).$$

For instance, if $X = \mathbb{P}_{\mathbb{C}}^N$, $E = \mathcal{O}(d_1) \oplus \ldots \oplus \mathcal{O}(d_n)$, $d_i \geq 1$, then the set of zeros of a section transverse to the zero section is simply the (smooth) intersection of n hypersurfaces of degrees $d_1, \ldots, d_n$ and the above formula for its Euler characteristic reads, with $d = N - n$,

$$\chi(Z) = \left(\prod_{i=1}^n d_i\right)\left[\sum_{i=0}^d (-1)^{d-i}\binom{N+1}{i} s_{d-i}(d_1, \ldots, d_n)\right],$$

where $s_{d-i}(-)$ is the $(d-i)^{th}$ complete (homogeneous) symmetric polynomial.

EXAMPLE. For $N = 5$, $n = 2$ and $d = 3$,

$$\chi(Z) = d_1 d_2\left[-(d_1^3 + d_2^3 + d_1^2 d_2 + d_1 d_2^2) + 6(d_1^2 + d_2^2 + d_1 d_2) - 15(d_1 + d_2) + 20\right].$$

From now on, let $\varphi : F \to E$ be a (holomorphic) morphism of vector bundles of ranks $m \geq n$ over a complex manifold X. To formulate the required transversality condition for φ, we consider the vector bundle $\text{Hom}(F, E)$ and the universal (or tautological) degeneracy loci $\mathbb{D}_i$ in it. The fiber of $\mathbb{D}_i$ over $x \in X$ is

$$\{f \in \text{Hom}\big(F(x), E(x)\big) \mid \text{rank}(f) \leq i\}.$$

Each $\mathbb{D}_i \setminus \mathbb{D}_{i-1}$ is nonsingular. For a given nonnegative integer r, we say that φ is r-**general** if the section s_φ of $\text{Hom}(F, E)$ induced by φ is transverse to all $\mathbb{D}_i \setminus \mathbb{D}_{i-1}$ for $i = 0, 1, \ldots, r$. This can be expressed equivalently by saying that all $D_i(\varphi) \setminus D_{i-1}(\varphi)$, $i = 0, 1, \ldots, r$, are nonsingular of (expected) codimension $(m-i)(n-i)$ in X.

EXERCISE. Prove the assertion contained in this last sentence. (See [P-P].)

Suppose that φ is r-general and $D_{r-1}(\varphi) = \emptyset$. Then $D := D_r(\varphi)$ is nonsingular and $\varphi|_D : F|_D \to E|_D$ is a map of constant rank r. In particular, we get two vector bundles on D:

$$K = \text{Ker}(\varphi|_D) \quad \text{and} \quad C = \text{Coker}(\varphi|_D)$$

of corresponding ranks $m - r$ and $n - r$. We claim that

(5.1) $N_{D/X} \cong K^\vee \otimes C.$

We first show that (5.1) holds for the normal bundle of $\mathbb{D}_i \setminus \mathbb{D}_{i-1}$ in $\text{Hom}(F, E)$ in the universal case, where K and C are the kernel and cokernel bundles of the tautological homomorphism on $\mathrm{Hom}(F, E)$. This last assertion is a consequence of the following statement. Let, temporarily, E and F denote two vector spaces of dimensions n and m and D_r the variety of all linear maps in $\mathrm{Hom}(F, E)$ of rank at most r. Pick $f \in D_r$ of rank r. Then the normal space to D_r in $\mathrm{Hom}(F, E)$ at f is canonically isomorphic to $\mathrm{Hom}\big(\mathrm{Ker}(f), \mathrm{Coker}(f)\big)$.

Here is a simple differential geometry argument for that.[1] Let f_t be a curve in D_r with the initial point $f_0 = f$. Choose a curve v_t in $\mathrm{Ker}(f_t) \subset F$. Then

$$f_t \cdot v_t = 0,$$

where f_t is viewed as a matrix, v_t as a column vector, and the dot denotes matrix multiplication. Differentiating w.r.t. t and evaluating at $t = 0$, we get

$$\dot{f}_0(v_0) + f(\dot{v}_0) = 0,$$

the dots denoting the corresponding derivatives. Hence $\dot{f}_0(\mathrm{Ker}(f)) \subset \mathrm{Im}(f)$, and consequently the tangent space to D_r in $\mathrm{Hom}(F, E)$ at f is contained in

$$\{h \in \mathrm{Hom}(F, E) \,:\, h(\mathrm{Ker}(f)) \subset \mathrm{Im}(f)\} =: V.$$

Finally, the equality

$$\dim \mathrm{Hom}\big(\mathrm{Ker}(f), \mathrm{Coker}(f)\big) = (m - r)(n - r)$$

together with the exact sequence

$$0 \to V \to \mathrm{Hom}(F, E) \to \mathrm{Hom}\big(\mathrm{Ker}(f), \mathrm{Coker}(f)\big) \to 0$$

show that the tangent space to D_r at f is V, and the normal space to D_r at f is isomorphic to $\mathrm{Hom}\big(\mathrm{Ker}(f), \mathrm{Coker}(f)\big)$, as claimed.

EXERCISE. Using the just proven property of the universal case, show (5.1).

Supposing now additionally that X is compact, and using (5.1), we will calculate $\chi(D)$. Denoting by $\iota : D \hookrightarrow X$ the inclusion, we get

(5.2)
$$\chi(D) = \int_D c(D) = \int_X \iota_* c(TX|_D - K^\vee \otimes C)$$
$$= \int_X \sum_p (-1)^p \iota_* s_p(K^\vee \otimes C) c(X).$$

To carry out the calculation, we need two formulas. The first expresses the total Segre class of the tensor product of two vector bundles A and B with ranks a and b such that $a \le b$:

(5.3) $$s(A \otimes B) = \sum D^{a;b}_{\lambda;\mu}\, s_\lambda(A)\, s_\mu(B),$$

where the sum is over pairs of partitions $\lambda = (\lambda_1, \ldots, \lambda_a)$, $\mu = (\mu_1, \ldots, \mu_a)$ and $D^{a;b}_{\lambda;\mu}$ is the binomial determinant defined by

$$D^{a;b}_{\lambda;\mu} = \det\left[\binom{\lambda_i + \mu_j + a + b - i - j}{\lambda_i + a - i}\right]_{1 \le i, j \le a}$$

(see [La-La-T]).

---

[1] We learned this argument from a preliminary version of [H-T1]. See also [G-G, p.145].

EXAMPLE. 1) If $a = \mathrm{rank}(A) = 1$, then

$$s(A \otimes B) = \sum_{l,m} \binom{l+m+b-1}{l} c_1(A)^l s_m(B).$$

2) If $a = b = 2$, then $s(A \otimes B)$ is equal to

$$\sum \begin{vmatrix} \binom{\lambda_1+\mu_1+2}{\lambda_1+1} & \binom{\lambda_1+\mu_2+1}{\lambda_1+1} \\ \binom{\lambda_2+\mu_1+1}{\lambda_2} & \binom{\lambda_2+\mu_2}{\lambda_2} \end{vmatrix} s_{\lambda_1,\lambda_2}(A) s_{\mu_1,\mu_2}(B),$$

the sum over pairs of nonnegative integers $\lambda_1 \geq \lambda_2$, $\mu_1 \geq \mu_2$.

We need also a formula which evaluates the Chern numbers of the kernel or cokernel bundles, or more generally $\iota_*\big(s_\lambda(C) \cdot s_\mu(K)\big)$ as polynomials in the Chern classes of E and F. Using the notation of Chapter 3, if $l(\lambda) \leq n-r$ and $l(\tilde{\mu}) \leq m-r$, then

(5.4) $$\iota_*\big(s_\lambda(C) \cdot s_\mu(-K)\big) = s_{((m-r)^{n-r}+\lambda),\mu}(E - F),$$

(see [P1, §5]).

EXAMPLE. Let $m = 6$, $n = 5$, $r = 2$. We have

$$\iota_*\big(s_{4,3,1}(C) \cdot s_{3,2}(-K)\big) = s_{8,7,5,3,2}(E - F).$$

This is displayed diagrammatically:

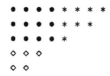

Formula (5.4) can be deduced from Proposition 1 from Chapter 4. Using the notation of that proposition, let s be the section of H induced by φ. Let Y be the variety of zeros of s. Denote by ρ the restriction of τ to Y. We have a commutative diagram

$$\begin{array}{ccc} Y & \xrightarrow{\jmath} & \mathbb{G} \\ \rho \downarrow & & \downarrow \tau \\ D & \xrightarrow{\iota} & X \end{array}$$

where $\jmath$ denotes the inclusion. Since φ is r-general and $D_{r-1}(\varphi) = \emptyset$, ρ establishes an isomorphism $Y \cong D$. This isomorphism allows one to identify K with $\jmath^*(S_F)$, C with $\jmath^*(Q_E)$ and ι with $\tau \circ \jmath$. Then (4.11) yields easily (5.4). (See [P1, Lemma 5.1].)

Substituting (5.3) and (5.4) to (5.2) and taking into account that

$$s_\mu(K^\vee) = (-1)^{|\mu|} s_{\tilde{\mu}}(-K),$$

we finally infer that

$$\chi(D) = \int_X P_r \cdot c(X),$$

where

$$P_r = P_r(E, F) := \sum_{\lambda, \mu} (-1)^{|\lambda| + |\mu|} D_{\lambda;\mu}^{n-r;m-r} s_{((m-r)^{n-r}+\lambda),\tilde\mu}(E - F)$$

and the sum is over all pairs of partitions $\lambda = (\lambda_1, \ldots, \lambda_{n-r})$, $\mu = (\mu_1, \ldots, \mu_{n-r})$. Set

(5.5) $$\Psi(r) := P_r \cdot c(X),$$

so that $\chi(D) = \int_X \Psi(r)$.

EXAMPLE. The Euler characteristic of a smooth determinantal curve $D_r(\varphi)$ (that is, $\dim(X) = (m - r)(n - r) + 1$, $\dim(D_r(\varphi)) = 1$ and $D_{r-1}(\varphi) = \emptyset$) is equal to

$$s_{(m-r)^{n-r}}(E - F)c_1(X) - (n - r)s_{(m-r)^{n-r},1}(E - F) - (m - r)s_{(m-r)^{n-r}+(1)}(E - F).$$

We now want to drop the assumption $D_{r-1}(\varphi) = \emptyset$. In this case $D_r(\varphi)$ is singular and it is a natural idea to try to compute first the Euler characteristic of a desingularization of $D_r = D_r(\varphi)$. To this end, we will work with the geometric construction already used in Chapter 4. Let us denote by Z_r the scheme of zeros of the homomorphism

$$F_{\mathcal{G}} \xrightarrow{\varphi_{\mathcal{G}}} E_{\mathcal{G}} \xrightarrow{\text{can}} Q,$$

contained in the total space of the Grassmann bundle $\pi = \pi_r : \mathcal{G} = G_r E \to X$.

LEMMA 1. *If φ is r-general then the section of $\text{Hom}(F_{\mathcal{G}}, Q)|_{Z_r}$ is transverse to the zero section of this bundle, and, consequently, the variety Z_r is nonsingular.*

This follows from the following result. Let $\alpha : A \to B$ be a surjective morphism of vector bundles. If a section s of A is transverse to the zero section of A, then the section $\alpha \circ s$ of B is transverse to the zero section of B. (See [P-P].)

Moreover, we have

$$N_{Z_r/\mathcal{G}} = F_{\mathcal{G}}^\vee \otimes Q|_{Z_r}.$$

Then using the Gauss-Bonnet formula and a well-known expression

$$[T\mathcal{G}] = [\pi^* TX] + [S^\vee \otimes Q]$$

in $K(\mathcal{G})$ (see e.g. [F1, App. B.5]), we get

$$\chi(Z_r) = \int_X \pi_* j_* c(Z_r),$$

where

$$\pi_* j_* c(Z_r) = \pi_* \big(c_{\text{top}}(F_{\mathcal{G}}^\vee \otimes Q) \, c(S^\vee \otimes Q - F_{\mathcal{G}}^\vee \otimes Q) \big) c(X).$$

Observe at this point that at least in theory we have the tools to express $\pi_* j_* c(Z_r)$ as a polynomial in the Chern classes of E, F and X. These tools are: Schur polynomial decompositions of $s(A \otimes B)$, $c(A \otimes B)$ (for the latter, see e.g [M2, p.67]), the Littlewood-Richardson rule and the push-forward formula. However, when one starts to compute the right-hand side directly just by using these tools, then it quickly becomes unwieldy, and the formulas contain many cancellations. Therefore we look for an alternative way of computing to avoid these difficulties. Let us denote

$$P_r' = P_r'(E, F) := \pi_* \big(c_{\text{top}}(F_{\mathcal{G}}^\vee \otimes Q) \, c(S^\vee \otimes Q - F_{\mathcal{G}}^\vee \otimes Q) \big).$$

We will show:

PROPOSITION. $P_r' = P_r$.

By its construction, the element P_r' is a specialization of a polynomial universally supported on the rth degeneracy locus with $c_i = c_i(E)$, $c_j' = c_j(F)$. As we recall from the previous chapter, we can write P_r' in the form

$$P_r' = \sum_\lambda a_\lambda(E) \, s_\lambda(E - F),$$

where the coefficients $a_\lambda(E)$ depend only on the Chern classes of E and are independent of the Chern classes of F. Then, analyzing when in P_r' an element $s_\mu(F)$ appears nontrivially (with the help of the formula (5.3) for the Segre class of the tensor product and the Littlewood-Richardson rule), one shows that $a_\lambda(E) \neq 0$ only if $\lambda \not\supset ((m - r + 1)^{n-r+1})$. For an algebro-combinatorial argument proving this last assertion, we refer the reader to [P-P].

This suggests that P_r' is *not* universally supported on the $(r - 1)^{th}$ degeneracy locus, and if so, in our calculation of $\chi(Z_r)$ we could "morally" assume that $D_{r-1}(\varphi) = \emptyset$, whence $D_r(\varphi) \cong Z_r$ and

$$\chi(Z_r) = \int_X \Psi(r),$$

(see (5.5) for the definition of $\Psi(r)$). It turns out that this speculation can be converted into a strict argument. The reader will find all details in [P-P] but a rough idea can be explained easily as follows.

Consider a commutative diagram of *Borel-Moore homology groups* associated with the *generic situation* used to prove the inclusion $\mathcal{P}_r \subset \mathcal{I}_r$ in Chapter 4.

$$
\begin{array}{ccc}
H_*(Z_r) & & \\
\downarrow {\scriptstyle \eta_*} & & \\
H_*(D_r) & \xrightarrow{k^*} & H_*(D_r \smallsetminus D_{r-1}) \\
\downarrow {\scriptstyle \iota_*} & & \downarrow \\
H_*(X) & \xrightarrow{l^*} & H_*(X \smallsetminus D_{r-1})
\end{array}
$$

where $D_r = D_r(\varphi)$, $k : D_r \setminus D_{r-1} \to D_r$ and $l : X \setminus D_{r-1} \to X$ are the inclusions. One has

$$k^* \eta_* \, c(Z_r) = c(-K^\vee \otimes C)(\iota k)^* \, c(X).$$

One constructs an element $a \in H_*(D_r)$ such that $\Psi(r) = \iota_*(a)$ and

$$k^*(a) = c(-K^\vee \otimes C)(\iota k)^* \, c(X).$$

Indeed, using Proposition 1 from Chapter 4 and the notation used in the sketch of the proof of (5.4), one can show that the element a of $H_*(D_r)$ defined by

$$a := \rho_* \Big(c\big(-(S_F^\vee \otimes Q_E)|_Y \big)(\tau_j)^* c(X) \Big),$$

satisfies the above two properties (cf. [P-P]). This implies that

$$l^* \big(\Psi(r) - \pi_* j_* \, c(Z_r) \big) = 0,$$

and consequently,

$$l^*(P_r - P_r') = 0.$$

Invoking the exact sequence of *Borel-Moore homology groups*

$$H_*(D_{r-1}) \xrightarrow{\bar{\iota}_*} H_*(X) \xrightarrow{l^*} H_*(X \setminus D_{r-1}),$$

where $\bar{\iota} : D_{r-1} \to X$ is the inclusion, we infer that $P_r' - P_r$ is a polynomial universally supported on the $(r-1)^{th}$ degeneracy locus specialized with $c_i = c_i(E)$ and $c_j' = c_j(F)$. But $P_r' - P_r$ is a $\mathbb{Z}[c_*(E)]$-combination of the $s_\lambda(E - F)$'s where all partitions λ satisfy the condition

$$\lambda \not\supset \big((m - r + 1)^{n - r + 1} \big)$$

and we know by the proposition in Section 4.3 that this forces

$$P_r' - P_r = 0,$$

as needed. For more details, we refer the reader to [P-P].

Supposing that we know $\chi(Z_r)$, we want now to write a system of linear equations relating the unknown quantities $\chi(D_r)$ with known quantities $\chi(Z_r)$. For that we need the following general result.

LEMMA 2. *If Y is a closed algebraic subset of a complex variety X, then*

$$\chi(X) = \chi(Y) + \chi(X \setminus Y).$$

A proof of this result can be found in [F8, p.142].

Consider the stratification of D_r by rank with the strata $D_k \setminus D_{k-1}$, $k = 0, 1, \ldots, r$. Set $Z_r^k := \pi_r^{-1}(D_k)$. Then $Z_r^k \setminus Z_r^{k-1}$ is a Grassmann bundle on

$D_k \setminus D_{k-1}$ with fiber the Grassmann manifold $G_{r-k}(\mathbb{C}^{n-k})$. Using the lemma, we obtain

$$\chi(Z_r) = \sum_{k=0}^{r} \chi(Z_r^k \setminus Z_r^{k-1}) = \sum_{k=0}^{r} \binom{n-k}{r-k} [\chi(D_k) - \chi(D_{k-1})]$$

$$= \sum_{k=0}^{r} \binom{n-k-1}{r-k} \chi(D_k).$$

Assuming that φ is r-general, we have

$$\chi(Z_i) = \int_X \Psi(i)$$

for $i \leq r$. By varying r, the above equation gives us a system of linear equations with the $\chi(D_r)$'s as unknown quantities, and possessing a triangular matrix of coefficients. Solving this system, we arrive at the final answer:

$$\chi(D_r(\varphi)) = \int_X \sum_{k=0}^{r} (-1)^k \binom{n-r+k-1}{k} \Psi(r-k).$$

In [P-P], a similar argument was used to compute the Chern-Schwartz-MacPherson class of the degeneracy loci (the degree of the zeroth component of this class is equal to the Euler characteristic $\chi(D_r(\varphi))$).

In [H-T2] and [P1, § 5], some algorithms for the computation of the Chern numbers of smooth degeneracy loci were given.

The description of the ideal of all polynomials universally supported on the rth degeneracy locus given in Chapter 4 can be also used to compute the homology of degeneracy loci. We refer the reader to [P1, § 4] for the computation of the Chow groups of determinantal varieties (over any base field) and their generalizations, along these lines.

Section 5.2 Brill-Noether loci in Jacobians

Let us sketch an application of this formula to Brill-Noether loci in Jacobians.

Recall that for a smooth curve C of genus g (over $\mathbb{C}$), the Brill-Noether locus

$$W_d^r(C) = \{L \in \mathrm{Pic}^d(C) : h^0(C, L) \geq r+1 \}$$

for $d \geq 1, r \geq 0$ and $g - d + r > 0$, parametrizes all complete degree d linear systems of dimension at least r. The set $W_d^r(C)$ inherits a natural scheme structure of a degeneracy locus defined as follows. Let $\mathcal{L}$ be a representative of the Poincaré line bundle on $\mathrm{Pic}^d(C) \times C$. Denote by

$$\nu : \mathrm{Pic}^d(C) \times C \to \mathrm{Pic}^d(C) \quad \text{and} \quad p : \mathrm{Pic}^d(C) \times C \to C$$

the corresponding projections. Fix an effective divisor of sufficiently positive degree m ($m \geq 2g - d + 1$ is sufficient). Write $\mathcal{L}(\pm D)$ instead of $\mathcal{L} \otimes p^* \mathcal{O}_C(\pm D)$. It follows from the Riemann-Roch theorem that $\mathcal{R}^1\nu_*\mathcal{L}(D) = 0$, thus $\nu_*\mathcal{L}(D)$ is locally free of rank $m + d - g + 1$. Taking the direct images of the short exact sequence

$$0 \to \mathcal{L} \to \mathcal{L}(D) \to \mathcal{L}(D)/\mathcal{L} \to 0,$$

we get the exact sequence

(5.6) $$0 \to \nu_*\mathcal{L} \to \nu_*\mathcal{L}(D) \to \nu_*\big(\mathcal{L}(D)/\mathcal{L}\big) \to \mathcal{R}^1\nu_*\mathcal{L} \to 0.$$

Of course, the scheme $W_d^r(C)$ should be supported on the set

$$\{L \in \mathrm{Pic}^d(C) : \dim(\nu_*\mathcal{L})_{\{L\}} \geq r + 1\}.$$

Since rank $\nu_*\big(\mathcal{L}(D)/\mathcal{L}\big) = m$, this last condition is equivalent to

$$\dim(\mathcal{R}^1\nu_*\mathcal{L})_{\{L\}} \geq g - d + r,$$

and we see that it is natural to treat $W_d^r(C)$ as the following degeneracy locus

$$W_d^r(C) = D_{m-g+d-r}(\varphi'),$$

where

$$\varphi' : F' = \nu_*\mathcal{L}(D) \to E' = \nu_*\big(\mathcal{L}(D)/\mathcal{L}\big)$$

is a homomorphism of vector bundles of ranks $m + d - g + 1$ and m respectively involved in the sequence (5.6). It turns out that this scheme structure on $W_d^r(C)$ doesn't depend on a choice of $\mathcal{L}$ and D (see [A-C-G-H, IV§3]). Also, it is shown in [A-C-G-H, VII§2] that by a suitable shift one can present $W_d^r(C)$ as $D_{m-g+d-r}(\varphi)$ where

$$\varphi : F = \nu_*\mathcal{L} \to E = \nu_*\big(\mathcal{L}/\mathcal{L}(-D)\big)$$

is a morphism of vector bundles of the same ranks $m+d-g+1$ and m on $\mathrm{Pic}^{m+d}(C)$. By normalizing $\mathcal{L}$ in such a way that

$$\mathcal{L}|_{\mathrm{Pic}^{m+d}(C)\times\{c\}} \in \mathrm{Pic}^0\big(\mathrm{Pic}^{m+d}(C)\big)$$

for a point $c \in C$, the Chern classes of E and F are pretty simple:

$$c(-F) = 1 + \theta + \frac{1}{2!}\theta^2 + \frac{1}{3!}\theta^3 + \dots \quad \text{and} \quad c(E) = 0,$$

where θ is the class of the theta divisor on $\mathrm{Pic}^{m+d}(C)$. The derivation of these identities uses the Poincaré formula and can be found in [A-C-G-H, VII§4].[2] Recall that θ defines a principal polarization of $\mathrm{Pic}^{m+d}(C)$ and $\int \theta^g = g!$.

Let

$$\rho := g - (r + 1)(g - d + r)$$

be the Brill-Noether number which is the expected dimension of $W_d^r(C)$. Let us assume from now on that C is *general* in the sense of moduli. Then $\dim(W_d^r(C)) = \rho$. If $\rho > 0$ then $W_d^r(C)$ is irreducible, and the singular locus of $W_d^r(C)$ equals $W_d^{r+1}(C)$ (see [A-C-G-H, V] and the references therein). Thus we are in position to apply the formula for the Euler characteristic of a degeneracy locus but we need the following simple result.

---

[2]These formulas are only true in cohomology or numerical equivalence rings, not Chow rings.

LEMMA. *Under the specialization* $c_i = \frac{1}{i!}$, $\Delta_\lambda(c) = \Delta_{\tilde\lambda}(c)$ *becomes* $\frac{1}{h(\lambda)}$, *where* $h(\lambda)$ *is the product of the hook lengths of all boxes in* D_λ.

We illustrate by the following picture what we understand by the *hook length* of a box.

The hook length of the black box in the picture is 6. In general, the **hook length of the box** $(i,j) \in D_\lambda$ (matrix coordinates) is

$$\lambda_i - i + \tilde\lambda_j - j + 1.$$

For example, the hook-lengths of the boxes in the depicted diagram are:

$$
\begin{array}{cccccc}
10 & 8 & 7 & 5 & 3 & 1 \\
8 & 6 & 5 & 3 & 1 \\
6 & 4 & 3 & 1 \\
4 & 2 & 1 \\
1
\end{array}
$$

For a proof of the lemma, we refer the reader to [M2, p.46].

EXERCISE. Prove that

$$h\big((r+1)^{g-d+r}\big) = \prod_{i=0}^{r} \frac{(g-d+r+i)!}{i!}.$$

Denote by $\Phi(g,d,r)$ the number

$$\Phi(g,d,r) := \int_{\mathrm{Pic}^{m+d} C} \Psi(m - g + d - r)$$

$$= (-1)^p g! \sum D_{\lambda;\mu}^{r+1;g-d+r} / h\big(((r+1)^{g-d+r} + \lambda), \tilde\mu\big),$$

where the sum is over pairs of partitions λ, μ with $l(\lambda)$ and $l(\mu)$ not greater than the minimum of $r+1$ and $g-d+r$. Using the lemma, the final result, for a general curve and $\rho \ge 0$, reads:

$$\chi\big(W_d^r(C)\big) = \sum_{k \ge r} (-1)^{k-r} \binom{k}{k-r} \Phi(g,d,k),$$

and implies immediately the following corollary.

Suppose that $g \not\equiv d \pmod 2$. Then $\chi\big(W_d^r(C)\big) < 0$ (resp. $\chi\big(W_d^r(C)\big) > 0$) iff $g \equiv r$ (resp. $g \not\equiv r$) (mod 2).

For example, if $\rho = 0$, then

$$\Phi(g,d,r) = \mathrm{card}(W_d^r(C)) = g!/h\big((r+1)^{g-d+r}\big)$$

is a classical formula of Castelnuovo.

For more details and similar computations of the Euler characteristics of some related objects in Brill-Noether theory, we refer the reader to [P-P].

Section 5.3 Morphisms with symmetries

We conclude this chapter by sketching some analogs of the results from Section 5.1 for symmetric and skew-symmetric bundle maps. Throughout this section, X will denote a complex manifold.

EXERCISE. Fix a nonnegative integer r. Suppose that for a symmetric morphism $\varphi : E^\vee \to E$ over X, the induced section $s_\varphi : X \to S^2 E$ is transverse to $\mathbb{D}_r^s \smallsetminus \mathbb{D}_{r-1}^s$. (Here, $\mathbb{D}_r^s \subset S^2 E$ is the tautological degeneracy locus whose fiber over $x \in X$ consists of all symmetric linear maps $E(x)^\vee \to E(x)$ of rank at most r). Show that the normal bundle of $D_r(\varphi) \smallsetminus D_{r-1}(\varphi)$ in $X \smallsetminus D_{r-1}(\varphi)$ is isomorphic to $S^2 C$, where C denotes the cokernel bundle of φ restricted to $D_r(\varphi) \smallsetminus D_{r-1}(\varphi)$.

Fix now an even nonnegative integer r. Suppose that for a skew-symmetric morphism $\varphi : E^\vee \to E$ over X, the induced section $s_\varphi : X \to \wedge^2 E$ is transverse to $\mathbb{D}_r^{ss} \smallsetminus \mathbb{D}_{r-2}^{ss}$. (Here, $\mathbb{D}_r^{ss} \subset \wedge^2 E$ is the tautological degeneracy locus whose fiber over $x \in X$ consists of all skew-symmetric linear maps $E(x)^\vee \to E(x)$ of rank at most r). Show that the normal bundle of $D_r(\varphi) \smallsetminus D_{r-2}(\varphi)$ in $X \smallsetminus D_{r-2}(\varphi)$ is isomorphic to $\wedge^2 C$, where C denotes the cokernel bundle of φ restricted to $D_r(\varphi) \smallsetminus D_{r-2}(\varphi)$.

The Chern numbers of the cokernel bundle C can be expressed in terms of Schur Q- and P-polynomials. Let X be compact. Keeping the assumption on φ from the exercise, and supposing additionally that $D_{r-1}(\varphi) = \emptyset$ for symmetric φ (resp. $D_{r-2}(\varphi) = \emptyset$ for skew-symmetric φ), the formulas in question are: in the symmetric case,

$$\iota_* s_\lambda(C) = Q_{\rho(n-r)+\lambda}(E);$$

and in the skew-symmetric case,

$$\iota_* s_\lambda(C) = P_{\rho(n-r-1)+\lambda}(E),$$

where $\iota : D_r(\varphi) \to X$ denotes the inclusion. For details, see [P1, § 7].

Formulas for the Segre classes of $S^2 E$ and $\wedge^2 E$, which are needed to compute the Euler characteristics of symmetric and skew-symmetric degeneracy loci have the following form. The total Segre class of the second symmetric power $S^2 E$ is given by

$$s(S^2 E) = \sum ((\lambda + \rho(n-1)))\, s_\lambda(E),$$

where the sum is over all partitions λ and the definition of $((\mu))$, for $\mu = (\mu_1 > \ldots > \mu_n \geq 0)$, is as follows. If n is even, define $((\mu))$ to be the Pfaffian of the $n \times n$ skew-symmetric matrix $(a_{i,j})$ where for $i < j$,

$$a_{i,j} = \sum \binom{\mu_i + \mu_j}{m} \qquad \text{(the sum over } \mu_j < m \leq \mu_i),$$

and if n is odd, then

$$((\mu)) := \sum_{i=1}^{n} (-1)^{i-1} 2^{\mu_i} ((\mu_1, \ldots, \hat{\mu}_i, \ldots, \mu_n)),$$

where "^" indicates the omission.

The total Segre class of the second exterior power $\wedge^2 E$ is given by

$$s(\wedge^2 E) = \sum \, [\lambda + \rho(n-1)] \; s_\lambda(E),$$

where the sum is over all partitions λ and the definition of $[\mu]$, for $\mu = (\mu_1 > \ldots > \mu_n \geq 0)$ is as follows. If n is even, define $[\mu]$ to be the Pfaffian of the $n \times n$ skew-symmetric matrix

$$\left((\mu_i + \mu_j - 1)!(\mu_i - \mu_j)/\mu_i! \mu_j! \right)_{1 \leq i,j \leq n} \; ;$$

if n is odd then $[\mu] = 0$ unless $\mu_n = 0$ where

$$[\mu] := [(\mu_1, \ldots, \mu_{n-1})].$$

For details, see [P1, §7] and [La-La-T].

Let X be compact and φ holomorphic. The formulas for the Euler characteristic of $D_r(\varphi)$, where we additionally assume that the preceding degeneracy locus is empty, have the following form. If φ is symmetric then

$$\chi(D_r(\varphi)) = \int_X \sum (-1)^{|\lambda|} ((\lambda + \rho(n-r-1))) \; Q_{\rho(n-r)+\lambda}(E)c(X),$$

the sum over all partitions λ of length $\leq n-r$. If φ is skew-symmetric then

$$\chi(D_r(\varphi)) = \int_X \sum (-1)^{|\lambda|} \, [\lambda + \rho(n-r-1)] \; P_{\rho(n-r-1)+\lambda}(E)c(X),$$

the sum over all partitions λ of length $\leq n-r$. For details, see [P1, §7]. For instance, for determinantal curves these formulas read: in the symmetric case,

$$\chi(D_r(\varphi)) = Q_{\rho(n-r)}(E)c_1(X) - (n-r+1)Q_{\rho(n-r)+(1)}(E);$$

and in the skew-symmetric case,

$$\chi(D_r(\varphi)) = P_{\rho(n-r-1)}(E)c_1(X) - (n-r-1)P_{\rho(n-r-1)+(1)}(E).$$

PROBLEM. It would be valuable to extend these formulas to "sufficiently general" symmetric and skew-symmetric morphisms without the assumption that the preceding degeneracy locus be empty, as we have done in this chapter for morphisms without symmetries. Here, a symmetric morphism $\varphi : E^\vee \to E$ on X should be "sufficiently general" if the induced section $s_\varphi : X \to S^2 E$ is transverse to all $\mathbb{D}_i^s \setminus \mathbb{D}_{i-1}^s$, where $i = 0, 1, \ldots, r$. Similarly, a skew-symmetric morphism $\varphi : E^\vee \to E$ on X should be "sufficiently general" if the induced section $s_\varphi : X \to \wedge^2 E$ is transverse to all $\mathbb{D}_i^{ss} \setminus \mathbb{D}_{i-2}^{ss}$, where $i = 0, 2, \ldots, r$.

CHAPTER VI

FLAG BUNDLES AND DETERMINANTAL
FORMULAS FOR THE OTHER CLASSICAL GROUPS

Section 6.1 Flag and Schubert varieties for the other classical groups

The Schubert variety story has an analogue for each of the semisimple groups G, with G/B (B a Borel subgroup) playing the role of the flag variety, and G/P (P a parabolic subgroup) playing the role of a Grassmannian or partial flag variety. This was carried out in the original work of [B-G-G] and [D2]. In the case of the flag variety G/B, there are Schubert varieties X_w and Y_w for each w in the corresponding Weyl group W.

Our aim here is to work out the general situation in the "modern" setting, with vector bundles, since that is what is needed for applications to algebraic geometry. However, it is a good idea first to take at least a brief look at the original case (when the base variety is a point). Most of what we say here will be said again in greater generality in the next section of this chapter.

$(\mathbf{C}_n)$. We start with the symplectic case, $(\mathbf{C}_n)$, which is in most respects next simplest to the $(\mathbf{A}_n)$ case of the general (or special) linear group. Here one is given a vector space V with a nondegenerate skew-symmetric form, also called a symplectic form, denoted $\langle \, , \, \rangle$. It is a standard fact that V must have even dimension, and that V has a basis $e_1, \ldots, e_{2n}$ so that $\langle e_i, e_{2n+1-i} \rangle = 1$ for $1 \le i \le n$, so $\langle e_{2n+1-i}, e_i \rangle = -1$, and the other $\langle e_i, e_j \rangle$ vanish.

Recall that a subspace L of V is isotropic if the symplectic form vanishes on it, i.e., $\langle u, v \rangle = 0$ for all u and v in L. The maximum dimension of such a subspace is n. A **complete isotropic flag** $L_\bullet$ is a chain $0 \subset L_1 \subset \cdots \subset L_n = L$ of isotropic subspaces, which is the same thing as a maximal isotropic subspace L (called Lagrangian) together with a complete flag of subspaces of L. Such a flag can be completed to a complete flag in V by setting $L_{n+i} = L_{n-i}{}^\perp$ for $1 \le i \le n$. The **standard** isotropic flag $V_\bullet$ has V_i generated by the first i basic vectors, $1 \le i \le 2n$. The flag variety $\mathcal{F}$ is the set of all such complete isotropic flags.

Note that any line in V is isotropic, so choosing L_1 amounts to picking a point in $\mathbb{P}(V)$. Having chosen L_1, one looks for a line L_2/L_1 in $L_1{}^\perp/L_1$, and so on. From this construction one sees that the dimension of $\mathcal{F}$ is $2n - 1 + 2n - 3 + \cdots + 1 = n^2$. (This also recovers the fact that the maximal dimension of an isotropic subspace is n.)

There are several different ways to describe the Weyl groups of the classical groups. The one we follow here allows a uniform explicit description of the Schubert varieties for the four families, but it requires translation to that with barred permutations (see Chapter 7) used more commonly. Here we set

$$W = \{\, w \in S_{2n} : w(i) + w(2n + 1 - i) = 2n + 1 \text{ for all } i \,\}.$$

Note that w is determined by its values on the integers between 1 and n, so it can be written $w = w(1)\, w(2) \ldots w(n)$. The cardinality of W is $2^n \cdot n!$. For w in W, define its length $l(w)$ by the formula

$$l(w) = \#\{i < j \le n : w(i) > w(j)\} + \#\{i \le j \le n : w(i) + w(j) > 2n + 1\}.$$

The element w_0 of longest length takes i to $2n + 1 - i$ for all i; its length is n^2.

There are also several conventions for Schubert varieties. Here we will use one of the conventions for which X_w has dimension $l(w)$. Fix a complete isotropic flag $V_\bullet$, for example the standard one described above. Set

$$X_w = \{\, L_\bullet : \dim(L_p \cap V_q) \ge \#\{i \le p : w(i) \le q\} \text{ for } 1 \le p, q \le 2n \,\}.$$

In fact, the conditions on the L_p for $p \le n$ determine those on the L_p for $p > n$, so one only needs to give these conditions for $1 \le p \le n$. The dual notion, giving a variety Y_w of codimension $l(w)$, is obtained by setting $Y_w = X_{w_0 \cdot w}$, i.e.,

$$Y_w = \{\, L_\bullet : \dim(L_p \cap V_q) \ge \#\{i \le p : w(i) > 2n - q\} \text{ for } 1 \le p \le n, 1 \le q \le 2n \,\}.$$

As in the (A_n) case, these Schubert varieties are closures of Schubert cells (defined by the same conditions but with the dimensions required to be equalities instead of inequalities). From this one sees that their classes form an additive basis of the cohomology of $\mathcal{F}$. On $\mathcal{F}$ we have the universal flag

$$0 = E_0 \subset E_1 \subset E_2 \subset \cdots \subset E_n = E \subset V$$

of subbundles. Set $x_i = -c_1(E_i/E_{i-1})$ in $H^2(\mathcal{F})$, $1 \le i \le n$. The cohomology of $\mathcal{F}$ is generated as a ring by these x_i's, and with relations obtained by setting the elementary symmetric polynomials in their squares $x_1{}^2, \ldots, x_n{}^2$ equal to zero. (This is part of the general calculation of Borel, but can be seen here directly: the symplectic form makes E and V/E dual vector bundles, from which it follows that $c(E) \cdot c(E^\vee) = 1$.) The problem solved in [B-G-G] and [D2] is to write the classes of the Schubert varieties as polynomials in these variables x_i, and conversely.

It is a good idea to work out some of these Schubert cells and varieties explicitly for small n. Usually one can find formulas for them by ad hoc arguments. In general, one can describe a flag $L_\bullet$ by a sequence of vectors $v_1, \ldots, v_n$, where L_i is spanned by the first i of these vectors; as before, one writes these as the rows of a matrix, using the standard basis. In this way, one can see the identification of the Schubert cells with affine spaces.

For example, for $n = 2$, the flag variety has dimension 4. The large cell X_w° for $w = w_0 = 4\,3$ can be described by matrices

$$\begin{bmatrix} a & b & c & 1 \\ t & d & 1 & 0 \end{bmatrix}$$

with $t = b - cd$, the other four coordinates identifying this cell with $\mathbb{A}^4$. Consider the Schubert variety X_w with $w = 3\,1$, whose length is 2. Unravelling the definition, it says that $L = L_2$ is contained in V_3. The cell is described by a matrix $\begin{bmatrix} a & b & 1 & 0 \\ 1 & 0 & 0 & 0 \end{bmatrix}$. This variety X_w is described by the condition that the map $E_2 \to V/V_3$ vanishes, so it is given by the second Chern class of $E_2{}^\vee$, which is $x_1 \cdot x_2$.

EXERCISE. Describe the other Schubert varieties for $n = 2$, and find formulas for them.

In general, the Schubert cell X_w° can be described by row echelon matrices of size $n \times 2n$, with 1's in the spots $(i, w(i))$, and 0's to the right of and below these 1's. Among the other entries, those *opposite* a position where there is a 1 in a higher row are determined, and the others are free. For example, for $n = 4$ and $w = 6\,4\,2\,8$, so $l(w) = 3 + 6 = 9$, such matrices have the form

$$\begin{bmatrix} * & * & * & * & * & 1 & 0 & 0 \\ * & * & \bullet & 1 & 0 & 0 & 0 & 0 \\ * & 1 & 0 & 0 & 0 & 0 & 0 & 0 \\ * & 0 & \bullet & 0 & \bullet & 0 & \bullet & 1 \end{bmatrix},$$

where we denote the free entries with *'s, and the determined entries with 1's, 0's, and •'s. It is an exercise to see that the number of *'s given by this prescription is the length of w, thus identifying the cell with an affine space of dimension $l(w)$.

($\mathbf{B_n}$). For this case, corresponding to the odd orthogonal groups, one has a vector space V of dimension $2n + 1$, with a nondegenerate quadratic form. One normalization for this form is to take a basis $e_1, \ldots, e_{2n+1}$, such that $\langle e_i, e_j \rangle = 1$ if $i + j = 2n + 2$, and $\langle e_i, e_j \rangle = 0$ otherwise. The dimension of a maximal isotropic subspace is n. An isotropic flag $L_1 \subset \cdots \subset L_n = L$ can be completed by setting $L_{n+i} = L_{n+1-i}^\perp$ for $1 \le i \le n+1$. The isotropic flag bundle $\mathcal{F}$ is defined as before. This time choosing an isotropic line amounts to choosing a point in a quadric hypersurface in $\mathbb{P}(V)$. Realizing $\mathcal{F}$ as by a sequence of quadric bundles, one sees that its dimension is again $2n - 1 + 2n - 3 + \cdots = n^2$. With these conventions, the Weyl group is

$$W = \{ w \in S_{2n+1} : w(i) + w(2n + 2 - i) = 2n + 2 \text{ for all } i \}.$$

Since $w(n+1) = n+1$, w is again determined by its values on the integers between 1 and n; the order of W is still $2^n \cdot n!$. Set

$$l(w) = \#\{i < j \le n : w(i) > w(j)\} + \#\{i \le j \le n : w(i) + w(j) > 2n + 2\}.$$

The standard isotropic flag $V_\bullet$ again has V_i spanned by the first i vectors. The locus X_w can be defined by the same formula as in the case ($\mathbf{C_n}$). For the locus Y_w of codimension $l(w)$ in case ($\mathbf{B_n}$), the description is

$$Y_w = \{ L_\bullet : \dim(L_p \cap V_q) \ge \#\{i \le p : w(i) > 2n + 1 - q\}$$
$$\text{for } 1 \le p \le n, 1 \le q \le 2n \}.$$

One has a similar description for the Schubert cells, using row echelon matrices of size n by $2n + 1$. This time, however, there are dots opposite 1's that are above *or* in the same row. For example, for $n = 4$ and $w = 4\,7\,9\,2$, with $l(w) = 3 + 6 = 9$, the Schubert cell X_w° can be described by matrices of the form

$$\begin{bmatrix} * & * & * & 1 & 0 & 0 & 0 & 0 & 0 \\ * & * & \bullet & 0 & * & \bullet & 1 & 0 & 0 \\ \bullet & * & \bullet & 0 & * & \bullet & 0 & * & 1 \\ \bullet & 1 & 0 & 0 & 0 & 0 & 0 & 0 & 0 \end{bmatrix}.$$

This flag manifold carries a tautological flag $0 \subset E_1 \subset \cdots \subset E_n \subset V$. The cohomology has generators $x_i = -c_1(E_i/E_{i-1})$, $1 \leq i \leq n$, but only if 2 is invertible in the coefficient ring. These generators satisfy the same relations as in the symplectic case, for essentially the same reason. (See the next section for details.)

Consider the Schubert variety X_w for $n = 2$ and $w = 5\,2$. The condition on the flag is that $L_2 \cap V_2$ is at least 1-dimensional. This is the same as the condition that $L_2 \cap V_3$ is at least 1-dimensional, which means that the map from $E_2 \to V/V_3$ drops rank. We know that the formula for such a locus is $c_2(E_2{}^\vee) = x_1 \cdot x_2$. However, a calculation in local coordinates shows that the determinant of this mapping vanishes doubly on the Schubert variety. In fact, a neighborhood of a point of X_w° in $\mathcal{F}$ is described by matrices

$$\begin{bmatrix} t & a & b & c & 1 \\ u & 1 & d & v & 0 \end{bmatrix}, \qquad \text{with} \qquad a, b, c, d \qquad \text{general,}$$

and $t + ac + b^2 = 0$, $v + d^2 = 0$, and $u + c + bd + va = 0$. The condition that $E_2 \to V/V_3$ drops rank is defined by the equation $v = 0$, while the condition that $L \cap V_2$ is at least 1-dimensional is defined by the equation $d = 0$. Since $v = -d^2$, one sees that the locus defined by the condition $\mathrm{rank}(E_2 \to V/V_3) \leq 1$ cuts out X_w with multiplicity 2, so the correct formula is $[X_w] = \frac{1}{2}x_1 \cdot x_2$.

EXERCISE. Work out similar formulas for all the Schubert varieties in case $n = 2$.

$(\mathbf{D}_n)$. Finally, we have the case corresponding to even orthogonal groups. Here V is a vector space of dimension $2n$, with a nondegenerate quadratic form. We take a basis $e_1, \ldots, e_{2n}$, such that $\langle e_i, e_j \rangle = 1$ if $i + j = 2n + 1$, and $\langle e_i, e_j \rangle = 0$ otherwise. The standard isotropic flag $V_\bullet$ again has V_i spanned by the first i vectors, and the dimension of a maximal isotropic subspace is n. This time there are two families of maximal isotropic subspaces L, those such that the dimension of $L \cap V_n$ is congruent to n modulo 2, and those where it is not; in the first case we say that L is **in the same family** as V_n. The isotropic flag manifold $\mathcal{F}$ is defined as before, with the added requirement that $L = L_n$ is in the same family as V_n. Note that if L_{n-1} is isotropic, there is a unique isotropic L_n containing it that is in the same family as V_n, since there is are exactly two isotropic lines in $L_{n-1}{}^\perp/L_{n-1}$.

The dimension of the flag manifold is $2n - 2 + 2n - 4 + \cdots = n^2 - n$. The Weyl group is

$$W = \{\, w \in S_{2n} : w(i) + w(2n + 1 - i) = 2n + 1 \text{ for all } i, \text{ and}$$
$$\text{the number of } i \leq n \text{ such that } w(i) > n \text{ is even}\,\}.$$

(This *is* a subgroup of S_{2n}.) Again w is determined by its values on the integers between 1 and n; in fact, it is determined by the first $n - 1$ values. The order of W is $2^{n-1} \cdot n!$. This time

$$l(w) = \#\{i < j \leq n : w(i) > w(j)\} + \#\{i < j \leq n : w(i) + w(j) > 2n + 1\}.$$

The locus X_w can be described by the same formula as in the cases $(\mathbf{C}_n)$ and $(\mathbf{B}_n)$. For Y_w it becomes

$$Y_w = \{\, L_\bullet : \dim(L_p \cap V_q) \geq \#\{i \leq p : w(i) > 2n - q\}$$
$$\text{for } 1 \leq p \leq n - 1, 1 \leq q \leq 2n \,\}.$$

This time, however, these inequalities must be taken with more than a grain of salt. The right definition is that the Schubert variety is the *closure* of the corresponding locus X_w° or Y_w° on which there is equality of dimensions. Consider the case where $n = 2$. The two Schubert varieties X_w of dimension 1 are given by $w = 3\,4$ and $w = 2\,1$. The first says that L_1 is contained in V_3, and the second that L_1 is contained in V_2 (which implies that L_2 is equal to V_2). This would indicate that the second locus is contained in the first, which is absurd. The first locus is really the closure of the locus where L_1 is contained in V_3 *and* $L_1 \cap V_2 = 0$. We will often write formulas for degeneracy loci in the displayed form, but they always need to be interpreted in this way.

There is a similar matrix description of the cells, following the conventions of the (B_n) case, but this time one needs only show $n-1$ rows. The corresponding Schubert cells described in the preceding paragraph are described by matrices $[\,* \bullet 1\,0\,]$ and $[\,*1\,0\,0\,]$. For $n = 4$, and $w = 4\,6\,2\,8$, so $l(w) = 2 + 4 = 6$, these matrices have the form

$$\begin{bmatrix} * & * & * & 1 & 0 & 0 & 0 & 0 \\ * & * & \bullet & 0 & \bullet & 1 & 0 & 0 \\ * & 1 & 0 & 0 & 0 & 0 & 0 & 0 \end{bmatrix}.$$

The cohomology of the flag manifold has one new equation, that $x_1 \cdot \ldots \cdot x_n = 0$, where, again, $x_i = -c_1(E_i/E_{i-1})$, $1 \le i \le n$, with the E_i the tautological subbundles of V; this can be seen from the fact that any nonisotropic vector in V gives a nowhere vanishing section of V/E_n, so $c_n(V/E_n) = 0$. (See the next section for more on this.) Again, these classes generate the cohomology only if 2 is invertible in the coefficient ring.

EXERCISE. Work out the formulas for classes of Schubert varieties in case $n = 2$, and do a few for $n = 3$.

In [B-G-G] and [D], analogues of the operators ∂_i were defined in the general setting. We will discuss these for the classical cases in the next section. Here it is particularly important to note that these operators were originally defined on cohomology rings, not on polynomials. Finding analogues of Schubert polynomials for the other groups is a nontrivial problem that has been receiving a lot of attention recently.

Section 6.2 Flag bundles and degeneracy loci

Our goal here is to describe analogues of the theorem we saw in Chapter 2 for the other classical groups, and for general vector bundles. The universal case of this will be a family of varying flag varieties. In carrying out these generalizations more difficulties are encountered than one might have expected.

Here again important cases were considered by Giambelli. Let $A = (a_{i,j})$ be a symmetric $m \times m$ matrix of forms in variables $x_0, \ldots, x_N$, with the degree of $a_{i,j}$ equal to $p_i + p_j$ for some integers $p_1, \ldots, p_m$. For any $r < m$, one has the locus $V_r \subset \mathbb{P}^N$ on which the rank of A is at most r. Let $k = m - r$. The matrix A is never generic enough for the general assumptions in Chapter 1 to apply. This time, for generic forms $a_{i,j}$, (for $i \le j$) the locus V_r is irreducible, of codimension $d = \binom{k+1}{2}$, (assuming $d < N$). Giambelli gave a formula for the degree (assuming $d \leq N$):

$$(6.1) \qquad \deg(V_r) = 2^k \cdot \Delta_\rho(p_1, \ldots, p_m), \quad \text{where} \quad \rho = (k, k-1, \ldots, 2, 1).$$

Similarly if A is skew-symmetric, the expected and generic codimension is $\binom{k}{2}$, and Giambelli's formula is

$$(6.2) \qquad \deg(V_r) = \Delta_\rho(p_1, \ldots, p_m), \quad \text{where} \quad \rho = (k-1, k-2, \ldots, 2, 1).$$

Since we will see these partitions often, let us set, for any positive integer k, the partition $\rho(k)$ by
$$\rho(k) = (k, k-1, \ldots, 2, 1).$$

Modern ("Thom-Porteous") versions of these formulas were given by Harris and Tu [H-T1], and by Józefiak, Lascoux, and Pragacz [J-L-P], [P2]. For this, one has a vector bundle E on a nonsingular variety X, and a map $\varphi : E \to E^\vee$ that is either symmetric or skew-symmetric, i.e., $\varphi = \varphi^\vee$ or $\varphi = -\varphi^\vee$, as described in Chapter 4. Let
$$D_r(\varphi) = \{ x \in X : \text{rank}(\varphi(x)) \leq r \}.$$

In the symmetric case, $D_r(\varphi)$ is locally defined by the vanishing of all minors of size $r+1$. In the skew-symmetric case, r is even, and one uses the Pfaffians of skew-symmetric submatrices obtained by choosing the same $r+2$ rows and columns.

If φ is symmetric, the expected codimension of $D_r(\varphi)$ is $\binom{k+1}{2}$, where $k = \text{rank}(E) - r$, and their formula is

$$(6.3) \qquad [D_r(\varphi)] = 2^k \cdot \Delta_{\rho(k)}(c(E^\vee)).$$

In the skew-symmetric case, the expected codimension of $D_r(\varphi)$ is $\binom{k}{2}$, where $k = \text{rank}(E) - r$, and the formula is

$$(6.4) \qquad [D_r(\varphi)] = \Delta_{\rho(k-1)}(c(E^\vee)).$$

Just as the Giambelli-Thom-Porteous formula is a special case of the general degeneracy formula we saw in Chapter 2 for the case of the general linear group, one should expect these formulas to be special cases of formulas for the orthogonal and symplectic groups.

Our interest in these questions was sparked by a question asked by Joe Harris about a decade ago. Suppose one is given a vector bundle V of rank $2n$ on a nonsingular variety X, together with a quadratic form on V, i.e., a symmetric bilinear form $V \otimes V \to 1_X$ to the trivial bundle, that is everywhere nondegenerate. (Here we assume the ground field has characteristic different from 2.) A subbundle E is called **isotropic** if this form is identically zero on E. The maximal rank of an isotropic subbundle is n.

Let E and F be two isotropic subbundles of rank n. It is a basic fact about quadratic forms (cf. [Mu1] and the end of this section) that the dimension of the intersection $E(x) \cap F(x)$ of the fibers at any point is a constant modulo 2 (on any connected base). This is the analogue of the familiar fact for lines on a quadric surface: two lines that are in the same family are either disjoint or coincide, and two in opposite families always meet in one point. We say that E and F are **in the same family** if

$$\dim\big(E(x) \cap F(x)\big) \equiv n \mod 2,$$

and that they are **in opposite families** otherwise.

If E and F are in the same family, let k be an integer congruent to n modulo 2. If E and F are in opposite families, let k be congruent to $n-1$ modulo 2. In either case, let

$$D_k = \big\{\, x \in X : \dim\big(E(x) \cap F(x)\big) \geq k \,\big\}.$$

Assuming sufficient generality, this is a subvariety of codimension $\binom{k}{2}$. Harris's question was to find a formula for the class $[D_k]$ in the cohomology of X. Here is the answer we found:

$$(6.5) \qquad [D_k] = \Delta_{\rho(k-1)}(c), \qquad c = \tfrac{1}{2}\big(c(E^{\vee}) + c(F^{\vee})\big),$$

i.e., $c_i = \tfrac{1}{2}(-1)^i\big(c_i(E) + c_i(F)\big)$. This formula has the sort of determinant we might (and did) expect, but its entries were unexpected, being sums of Chern classes of bundles — not, for example, Chern classes of sums of bundles. This formula is easiest to interpret if one works in cohomology with rational coefficients, but the right side is actually an integral class, and the formula is valid with integer coefficients.

One can ask the same question when V has rank $2n+1$. In this case the maximal rank of an isotropic subbundle is again n, and there are no questions about families. For any $k \leq n$ one has the locus D_k as before. In this case, the formula is

$$(6.6) \qquad [D_k] = \Delta_{\rho(k)}(c), \qquad c = \tfrac{1}{2}\big(c(E^{\vee}) + c(F^{\vee})\big).$$

Similarly, if V has rank $2n$, and $V \otimes V \to 1_X$ is a skew-symmetric form, everywhere nondegenerate, then the maximal rank of an isotropic subbundle is n (in which case they are usually called **Lagrangian**). If E and F are two such subbundles, the expected codimension of the locus D_k in this case is $\binom{k+1}{2}$, and this time our formula is

$$(6.7) \qquad [D_k] = \Delta_{\rho(k)}(c), \qquad c = c(E^{\vee}) + c(F^{\vee}).$$

Note in this case that $c_0 = 2$.

Let us first see how the formulas of Harris-Tu and Józefiak-Lascoux-Pragacz follow. Given a map $\varphi \colon E \to E^{\vee}$ of vector bundles, take $V = E \oplus E^{\vee}$. On V there are canonical nondegenerate bilinear forms:

$$\langle (e_1, f_1), (e_2, f_2) \rangle = f_1(e_2) \pm f_2(e_1), \quad e_i \in E, f_i \in E^{\vee},$$

with the $+$ sign giving a symmetric form, and the $-$ sign a skew-symmetric form. Let $E_\varphi \subset V$ be the graph of φ, i.e., the set of vectors of the form $(e, \varphi(e))$. Now it is easy to check that φ is *symmetric* exactly when E_φ is isotropic for the canonical *skew-symmetric* form on V, and φ is *skew-symmetric* exactly when E_φ is isotropic for the canonical *symmetric* form on V. (Note the switch!) The locus $D_r(\varphi)$ then becomes the locus D_k for the maximal isotropic subbundles E_φ and $E_0 = E \oplus 0$, with $k = \text{rank}(E) - r$. Note that E_φ and E_0 are bundles isomorphic to E, so $c(E_\varphi{}^\vee) + c(E_0{}^\vee) = 2c(E^\vee)$. Formula (6.3) then follows from formula (6.7), and formula (6.4) follows from formula (6.5).

By now we should expect these formulas to be special cases of general formulas, one for each element w of the corresponding Weyl group. We shall see next that this is indeed the case.

($\mathbf{C_n}$). We start with the symplectic case. Here we are given a vector bundle V of rank $2n$ on a nonsingular variety X, with a symplectic form as above. We are given two isotropic subbundles E and F of rank n, and, in addition, each of these bundles has a complete filtration:

$$E_\bullet : \ 0 = E_0 \subset E_1 \subset E_2 \subset \cdots \subset E_n = E \subset V,$$
$$F_\bullet : \ 0 = F_0 \subset F_1 \subset F_2 \subset \cdots \subset F_n = F \subset V.$$

These filtrations can be extended to complete flags in V, by setting $E_{n+i} = E_{n-i}{}^\perp$ and $F_{n+i} = F_{n-i}{}^\perp$ for $1 \leq i \leq n$. As in the first section, set

$$W = \{\, w \in S_{2n} : w(i) + w(2n + 1 - i) = 2n + 1 \quad \text{for all } i \,\}.$$

The group W is generated by n involutions[1] $s_1, \ldots, s_n$, where s_i interchanges i and $i + 1$, and also (necessarily) $2n - i$ and $2n - i + 1$, for $1 \leq i \leq n - 1$, and s_n interchanges n and $n+1$. The length $l(w)$ defined in the first section of this chapter is the minimal number one needs to write w as a product of these generators. Set

$$X_w = \{\, x \in X : \dim\bigl(E_p(x) \cap F_q(x)\bigr) \geq$$
$$\#\{i \leq p : w(i) \leq q\} \ \forall\, p, q \leq 2n \,\}.$$

Define cohomology classes in $H^2(X)$:

$$x_i = -c_1(E_i/E_{i-1}), \qquad y_i = -c_1(F_{n+1-i}/F_{n-i}), \qquad 1 \leq i \leq n.$$

Define the operators ∂_i for $1 \leq i \leq n - 1$, on polynomials in variables $x_1, \ldots, x_n$ (with any other variables regarded as scalars) by the same formula (1.8) as in Chapter 1. Define the operator ∂_n by the formula

$$\partial_n(P) = \bigl(P(x_1, \ldots, x_n) - P(x_1, \ldots, x_{n-1}, -x_n)\bigr)/2x_n.$$

For any w in W, define an operator ∂_w by the formula

$$\partial_w = \partial_{i_1} \circ \cdots \circ \partial_{i_l} \qquad \text{if } w = s_{i_1} \cdot \ldots \cdot s_{i_l}, \quad l = l(w).$$

As before, these operators are independent of choice.

---

[1]Another convention, better suited for embedding the flag variety for n in that for $n + 1$, denotes these $s_1, \ldots, s_n$ by $s_{n-1}, \ldots, s_0$.

THEOREM (C_n). *If $E_\bullet$ and $F_\bullet$ are in sufficiently general position, then*

$$[X_w] = \partial_{w^{-1}} \Big(\prod_{i+j \leq n} (x_i - y_j) \cdot \Delta_{\rho(n)}(c) \Big),$$

where $c_i = e_i(x_1, \ldots, x_n) + e_i(y_1, \ldots, y_n)$.

For another formulation, set

$$Y_w = \big\{\, x \in X : \dim\big(E_p(x) \cap F_q(x)\big) \geq \#\{i \leq p : w(i) > 2n - q\} \ \forall\, p, q \leq 2n \,\big\}.$$

The theorem is equivalent to the formula

(6.9) $$[Y_w] = \partial_{i_l} \circ \cdots \circ \partial_{i_1} \Big(\prod_{i+j \leq n} (x_i - y_j) \cdot \Delta_{\rho(n)}(c) \Big).$$

where $w = w_0 \cdot s_{i_1} \cdot \ldots \cdot s_{i_l}$, with l minimal; here $w_0 = 2n \ldots n + 1$. Formula (6.7) follows from this formula in the case where $k = n$, as one sees by taking w with $w(i) = n + 1 - i$ for $1 \leq i \leq n$. The cases where $k < n$ should follow, but this is not so obvious. In fact, if the appropriate operator is applied to the polynomial in the theorem, one does *not* get the polynomial prescribed in (6.7), but only one that defines the same cohomology class. This is the beginning of something we will see more of: for the other classical groups, there are not unique polynomials with all the properties one would like for Schubert and double Schubert polynomials. In practice, so far at least, the formulas of (6.7) are proved by the same method as the theorem.

Again in this situation there is a universal situation. We suppose we are given V of rank $2n$ on X with a symplectic form, and we are given a Lagrangian subbundle F with a complete flag $F_\bullet$ in F. Let $\mathcal{F} \to X$ denote the flag bundle, whose fiber at x consists of all flags $L_1 \subset \cdots \subset L_n = L \subset V(x)$, where L is Lagrangian. On $\mathcal{F}$ we have the tautological flag of bundles $E_1 \subset \cdots \subset E_n = E \subset V$. Using this flag and the pullback of the given flag $F_\bullet$, we can define universal Schubert varieties that we will denote by $\mathcal{X}_w$, $w \in W$. It suffices to prove the formula for $\mathcal{X}_w$, since any other case is pulled back by a section $s : X \to \mathcal{F}$.

First, one computes the cohomology $H^*(\mathcal{F})$ as an algebra over $H^*(X)$. The answer here is straightforward:

$$H^*(\mathcal{F}) = H^*(X)[x_1, \ldots, x_n] \,/\, I,$$

where I is the ideal generated by

(6.10) $$e_i(x_1{}^2, \ldots, x_n{}^2) - e_i(y_1{}^2, \ldots, y_n{}^2), \quad 1 \leq i \leq n.$$

These equations are easy to see geometrically. In fact, the nondegenerate form makes the bundle F and V/F into dual bundles, and this means that $c(F) \cdot c(F^\vee) = c(V)$, which means that

$$c(V) = \prod_{i=1}^{n} (1 + y_i)(1 - y_i) = \prod_{i=1}^{n} (1 - y_i{}^2),$$

so $(-1)^i c_{2i}(V) = e_i(y_1{}^2, \ldots, y_n{}^2)$. Since the same holds for the bundle E on $\mathcal{F}$ in place of F, the equations follow. In fact, the monomials

$$(6.11) \qquad x_1{}^{i_1} \cdot \ldots \cdot x_n{}^{i_n} \cdot e_1(x)^{j_1} \cdot \ldots \cdot e_n(x)^{j_n}, \qquad i_\alpha \leq n - \alpha, \ j_\beta \leq 1,$$

form an additive basis for $H^*(\mathcal{F})$ as a module over $H^*(X)$. (The fact that these give a basis for each fiber, together with a Mayer-Vietoris argument, shows that they give a global basis, and from this it follows that the n given equations span I.)

For the proof of the theorem, again there are two steps. First, we must find the formula for the subvariety $\mathcal{X}_{\mathrm{id}}$ of largest codimension (corresponding to a point in the flag manifold), i.e., the locus of flags equal to the fixed flag $F_\bullet$. This is surprisingly more difficult than the (A_n) case. The problem is that we know of no vector bundle K of rank n^2 on $\mathcal{F}$ with a section that cuts out this locus. The idea for our proof is to first find a formula for the locus where E_1 is contained in F, which is given by a top Chern class. On this locus one has $E_1 \subset F \subset E_{2n-1}$, and one can find a formula for the locus where E_2 is contained in F. Continuing in this way, one arrives at a formula for the locus where $E = F$. Unfortunately, this formula looks little like the desired formula $\Delta_{\rho(n)}(c)$, and one has to do some algebra in the cohomology ring to prove this; for details see [F3]. (For another approach, see Chapter 7, [P-R5], and also Appendix E.)

The second step, to go from one Schubert variety $\mathcal{X}_w$ to the next, $\mathcal{X}_{w \cdot s_i}$, is essentially the same as in the (A_n) case discussed in Chapter 2.

$(\mathbf{B_n})$. The (B_n) case, where V is a bundle of rank $2n+1$, with a quadratic form, is quite similar. We describe the differences. One is given flags of isotropic bundles as before, with E and F of rank n. This time they are completed to complete flags by setting $E_{n+i} = E_{n+1-i}^\perp$ and $F_{n+i} = F_{n+1-i}^\perp$ for $1 \leq i \leq n+1$. The Weyl group is

$$W = \{\, w \in S_{2n+1} : w(i) + w(2n + 2 - i) = 2n + 2 \quad \text{for all } i \,\}.$$

The n involutions generating W are: s_i, that interchanges i and $i+1$, and $2n+1-i$ and $2n+2-i$, for $1 \leq i \leq n-1$, and s_n, that interchanges n and $n+2$. This time ∂_n is twice that for (C_n):

$$\partial_n(P) = \big(P(x_1, \ldots, x_n) - P(x_1, \ldots, x_{n-1}, -x_n)\big) / x_n.$$

The loci X_w, and the classes x_i and y_i are defined as before, and the theorem reads

THEOREM $(\mathbf{B_n})$. *The formula for the degeneracy locus X_w is*

$$[X_w] = \partial_{w^{-1}} \Big(\prod_{i+j \leq n} (x_i - y_j) \cdot \Delta_{\rho(n)}(c) \Big),$$

where $c_i = \frac{1}{2}\big(e_i(x_1, \ldots, x_n) + e_i(y_1, \ldots, y_n)\big)$.

If

$$Y_w = \big\{\, x \in X : \dim\big(E_p(x) \cap F_q(x)\big) \geq \#\{i \leq p : w(i) > 2n + 1 - q\}$$
$$\text{for } 1 \leq p \leq n, \ 1 \leq q \leq 2n \,\big\},$$

the theorem is equivalent to the formula

$$(6.12) \qquad [Y_w] = \partial_{i_l} \circ \cdots \circ \partial_{i_1} \Big(\prod_{i+j \leq n} (x_i - y_j) \cdot \Delta_{\rho(n)}(c) \Big),$$

where $w = w_0 \cdot s_{i_1} \cdot \ldots \cdot s_{i_l}$, with l minimal; here $w_0 = 2n + 1 \ldots n + 2$.

The presentation of the cohomology ring of $\mathcal{F}$ over that of X is the same as in the (C_n) case, except that, this time, it is only valid with coefficients in $\mathbb{Z}[\frac{1}{2}]$: the classes x_i do not generate the cohomology ring of $\mathcal{F}$ with integer coefficients. The fact that the relations are the same is not quite as obvious, however. Here the quadratic form makes $F_n = F$ and V/F_{n+1} dual bundles. It also makes F_{n+1}/F_n isomorphic to its dual, which only implies that twice its first Chern class vanishes. However, we have an isomorphism

$$\wedge^{2n+1}(V) \cong \wedge^n(F) \otimes (F_{n+1}/F_n) \otimes \wedge^n(V/F_{n+1}) \cong F_{n+1}/F_n,$$

which implies that $E_{n+1}/E_n \cong F_{n+1}/F_n$, and we therefore have the same equations $c(E) \cdot c(E^\vee) = c(F) \cdot c(F^\vee)$ as before.

Another new wrinkle does appear here. In the previous cases, the loci X_w that we have described set-theoretically were cut out by the obvious equations — corresponding to minors of matrices. This is no longer true for the symmetric loci. In fact, some of these equations cut out the corresponding loci with multiplicity two, as we saw in the preceding section. (For more about the equations of Schubert varieties, see [Lk-Sh].) This explains why the formula for ∂_n is twice what it was before. In this case, the simplest way to describe the loci X_w as algebraic subschemes is to take $\mathcal{X}_w$ in the universal case to be corresponding set with its reduced structure. Then any other locus X_w is equal to $s^{-1}(\mathcal{X}_w)$, for a section $s : X \to \mathcal{F}$, and this gives local equations for X_w. The rest of the story is quite similar to the (C_n) case.

(D_n). For the (D_n) case, one has V of rank $2n$, with a quadratic form, and rank n isotropic subbundles E and F with complete flags of subbundles as before. This time (in order for the corresponding flag manifold to be connected), we assume that E and F are in the same family. In this case it is not so obvious how to describe the cohomology ring of $\mathcal{F}$. There are the same equations (6.10), for the same reason, but these equations do not suffice. Some experimenting in low ranks led to the conjecture that E and F must always have equal top Chern classes, when they are in the same family, and that $c_n(E) = -c_n(F)$ if they are in opposite families (refining the equation $c_n(E)^2 = c_n(F)^2$ that we knew).

This conjecture was proved by Edidin and Graham [E-G], which provided a key to our making progress on this problem. (See the next section for more on this.) Hence, with the same x_i and y_i as before, and with coefficients in $\mathbb{Z}[\frac{1}{2}]$,

$$H^*(\mathcal{F}) = H^*(X)[x_1, \ldots, x_n] / I,$$

where I is the ideal generated by

$$(6.13) \qquad \begin{aligned} & e_i(x_1{}^2, \ldots, x_n{}^2) - e_i(y_1{}^2, \ldots, y_n{}^2), \quad 1 \leq i \leq n-1, \quad \text{and} \\ & e_n(x_1, \ldots, x_n) - e_n(y_1, \ldots, y_n). \end{aligned}$$

The monomials

$$(6.14) \qquad x_1{}^{i_1} \cdot \ldots \cdot x_n{}^{i_n} \cdot e_1(x)^{j_1} \cdot \ldots \cdot e_{n-1}(x)^{j_{n-1}}, \quad i_\alpha \le n - \alpha, \quad j_\beta \le 1,$$

form an additive basis for $H^*(\mathcal{F})$ as a module over $H^*(X)$.

The Weyl group is

$$W = \{\, w \in S_{2n} : w(i) + w(2n + 1 - i) = 2n + 1 \;\; \text{for all } i, \text{ and}$$
$$\text{the number of } i \le n \text{ such that } w(i) > n \text{ is even}\,\}.$$

The generators s_i are as in the (C_n) case for $1 \le i \le n - 1$, and s_n interchanges $n - 1$ and $n + 1$ and also interchanges n and $n + 2$. This time ∂_n is defined by the formula

$$\partial_n(P) = \big(P(x_1, \ldots, x_n) - P(x_1, \ldots, x_{n-2}, -x_n, -x_{n-1})\big) / (x_{n-1} + x_n).$$

The loci X_w are defined by the same formula. The theorem in this case reads

THEOREM (D_n). *The formula for the degeneracy locus X_w is*

$$[X_w] = \partial_{w^{-1}} \Big(\prod_{i+j \le n} (x_i - y_j) \cdot \Delta_{\rho(n-1)}(c) \Big),$$

where $c_i = \frac{1}{2}\big(e_i(x_1, \ldots, x_n) + e_i(y_1, \ldots, y_n)\big).$

Equivalently, if

$$Y_w = \big\{\, x \in X \; : \; \dim\big(E_p(x) \cap F_q(x)\big) \ge \#\{ i \le p : w(i) > 2n - q \}$$
$$\text{for } 1 \le p \le n - 1, \; 1 \le q \le 2n \,\big\},$$

then

$$(6.15) \qquad [Y_w] = \partial_{i_l} \circ \cdots \circ \partial_{i_1} \Big(\prod_{i+j \le n} (x_i - y_j) \cdot \Delta_{\rho(n-1)}(c) \Big),$$

where $w = w_0 \cdot s_{i_1} \cdot \ldots \cdot s_{i_l}$, with l minimal; the first $n-1$ values of w_0 are $2n \ldots n+2$.

$(\widetilde{D}_n)$. There is another case, that we denote by $(\widetilde{D}_n)$, with V as in the preceding case, but this time E and F are in opposite families. Set

$$\widetilde{W} = \{\, \widetilde{w} \in S_{2n} : \widetilde{w}(i) + \widetilde{w}(2n + 1 - i) = 2n + 1 \;\; \text{for all } i \text{ and}$$
$$\text{the number of } i \le n \text{ such that } \widetilde{w}(i) > n \text{ is odd}\,\}.$$

Note this time that $\widetilde{W}$ is not a subgroup of S_{2n}; there is a one-to-one correspondence between the Weyl group W for (D_n) and $\widetilde{W}$: given $w \in W$, define $\widetilde{w}$ by the formula $\widetilde{w}(i) = w(i)$ for $i \notin \{n, n + 1\}$, and $\widetilde{w}(i) = 2n + 1 - w(i)$ for $i \in \{n, n + 1\}$. This correspondence preserves length, with length for $\widetilde{W}$ defined by the same formula as in the (D_n) case. For each $\widetilde{w}$ in $\widetilde{W}$ there is a locus $X_{\widetilde{w}}$ defined by the same conditions as in the other cases: the dimension of $E_p(x) \cap F_q(x)$ is at least the cardinality of $\{i \leq p : \widetilde{w}(i) \leq q\}$ for all $1 \leq p, q \leq 2n$.

Given an isotropic flag $E_\bullet$, there is a unique isotropic flag $\widetilde{E}_\bullet$ defined by setting $\widetilde{E}_i = E_i$ for $i \leq n - 1$, defining $\widetilde{E}_n$ so that $E_{n-1}{}^\perp/E_{n-1}$ is the direct sum of E_n/E_{n-1} and $\widetilde{E}_n/E_{n-1}$. If E_n and F are in the opposite family, then $\widetilde{E}_n$ and F are in the same family, and $E_\bullet$ is in $X_{\widetilde{w}}$ exactly when $\widetilde{E}_\bullet$ is in X_w. We can therefore give $X_{\widetilde{w}}$ its precise definition and scheme-theoretic structure by letting $s : X \to \mathcal{F}$ be the section of the flag bundle $\mathcal{F}$ for (D_n) such that $s^*(U_\bullet) = \widetilde{E}_\bullet$, and defining $X_{\widetilde{w}}$ to be $s^{-1}(\mathcal{X}_w)$. The formula for $[X_{\widetilde{w}}]$ in the $(\widetilde{D}_n)$ case is the same as the formula for $[X_w]$ in the (D_n) case, with the only change that x_n is replaced by $-x_n$. The same formula as in the (D_n) case describes the degeneracy locus $Y_{\widetilde{w}}$, and one has the corresponding formula for it.

Once one has reached this stage, the proofs of the theorems for the (B_n) and (D_n) cases are quite similar to the (C_n) case.

Section 6.3 Integrality

To have formulas with integer coefficients, one uses the ideas of [E-G]. Consider the (B_n) case. Here one has the quadric bundle $\mathbb{Q}(V)$ of isotropic lines in V. This is a hypersurface in $\mathbb{P}(V)$, whose class $[\mathbb{Q}(V)]$ is $2H$, where $H = c_1(\mathcal{O}(1))$. With F a maximal isotropic subbundle as before, we have $\mathbb{P}(F) \subset \mathbb{Q}(V) \subset \mathbb{P}(V)$. If γ is the class of $\mathbb{P}(F)$ in $H^{2n}(\mathbb{Q}(V))$, and $h \in H^2(\mathbb{Q}(V))$ is the restriction of H, then the classes h^i and $h^i \cdot \gamma$ for $0 \leq i \leq n - 1$, give a basis for $H^*(\mathbb{Q}(V))$ over $H^*(X)$ (all now with integer coefficients). If $\mathcal{G}$ denotes the Grassmannian of isotropic n-planes in V, and E is the tautological subbundle on $\mathcal{G}$, then there is an equation

$$[\mathbb{P}(E)] - [\mathbb{P}(F_\mathcal{G})] = \sum_{i=1}^{n} H^{n-i} \cdot a_i,$$

for some classes a_i in $H^*(\mathcal{G})$. In fact,

(6.15) $$2 a_i = c_i(E^\vee) - c_i(F^\vee) = e_i(x) - e_i(y).$$

The polynomials $a_1{}^{j_1} \cdot \ldots \cdot a_n{}^{j_n}$, where $j_\alpha = 0$ or 1, form a basis for $H^*(\mathcal{G})$ over $H^*(X)$. It follows that the classes

$$x_1{}^{i_1} \cdot \ldots \cdot x_n{}^{i_n} \cdot a_1{}^{j_1} \cdot \ldots \cdot a_n{}^{j_n},$$

with $i_\alpha \leq n - \alpha$ and $j_\beta = 0$ or 1, form a basis for $H^*(\mathcal{F})$ over $H^*(X)$, all with integer coefficients. Note that the c_i that occur in the basic theorem are equal to $a_i + 2e_i(y)$, so the right sides of the formulas make sense with integer coefficients. The theorem is of the last section is then valid with integer coefficients. Essentially the same is true in the (D_n) case.

Section 6.4 Twisting by a line bundle

For more useful formulas, one must improve them to allow the quadratic or symplectic form to have values in an arbitrary line bundle L, not just in a trivial bundle, so the bilinear form is a map $V \otimes V \to L$. For example, this allows applications to skew-symmetric or symmetric maps $E \to E^{\vee} \otimes L$, cf. [H-T1], [P2]. This is what one sees with a skew-symmetric or symmetric matrix of forms on projective space, with $E = \oplus \mathcal{O}(-p_i)$, and $L = \mathcal{O}(m)$ for some m, so the entries are forms of degrees $p_i + p_j + m$. If m is even, one can deal without L by giving half of m to each p_i, but if m is odd, this is not possible.

In general, if $L \cong M^{\otimes 2}$ is the square of a line bundle M, one can replace V by $V \otimes M^{\vee}$, to reduce to the case considered before. This tensors each E_i and F_i by $M^{\vee}$, so adds $v = c_1(M)$ to each x_i and y_i. If one uses rational coefficients (or coefficients in $\mathbb{Z}[\frac{1}{2}]$), in fact, the same formulas are valid, once this change is made, with $v = \frac{1}{2}c_1(L)$. (One must also include v appropriately in the formula for ∂_n.) To prove this, one constructs a two-sheeted branched covering $\pi : X' \to X$, such that $\pi^*(L) = M^{\otimes 2}$ for some M. Since $\pi_* \circ \pi^*$ is multiplication by 2, formulas valid on X' imply their validity on X, provided one ignores 2-torsion.

If one wants a formula valid over $\mathbb{Z}$, however, one must argue differently. In fact, in spite of an assertion in [H-T1], Totaro showed that there *cannot* always be a map $\pi : X' \to X$, such that $\pi^*(L) = M^{\otimes 2}$ and such that π^* is an injection on integral cohomology. All the formulas we have discussed here, however, are valid as if this principle were true. To prove them, one can argue that, after appropriate modification, they can be pulled back from a situation in which there is no torsion at all in the cohomology (or Chow groups) — in which case the argument of the preceding paragraph is valid. For details of this and the other arguments of this chapter, see [F3].

$\tilde{P}$- AND $\tilde{Q}$-POLYNOMIAL FORMULAS
FOR OTHER CLASSICAL GROUPS

In this chapter, we will discuss a certain way of globalizing the formulas for fundamental classes of Schubert varieties in isotropic Grassmannians from Chapter 3. Let, for simplicity, X denote a nonsingular (complex) variety. Suppose that V is a vector bundle on X, endowed with an everywhere nondegenerate bilinear form ϕ. We will consider three cases:

(C_n) Lagrangian: $\operatorname{rank}(V) = 2n$ and ϕ is skew-symmetric;
(B_n) odd orthogonal: $\operatorname{rank}(V) = 2n + 1$ and ϕ is symmetric;
(D_n) even orthogonal: $\operatorname{rank}(V) = 2n$ and ϕ is symmetric.

Recall from the previous chapter the question raised by J. Harris (originally in the even orthogonal case):

Suppose that E and F are maximal isotropic subbundles of V. Find a polynomial in the Chern classes of E and F, which represents the locus

$$\{x \in X \,:\, \dim(E \cap F)(x) \geq k\},$$

where k is a fixed positive integer.

In this chapter, we will give a solution to this problem, considering, as in Chapter 6, a more general situation. Assume we have an isotropic flag

$$F_\bullet : 0 = F_0 \subset F_1 \subset F_2 \subset \ldots \subset F_n = F$$

with $\operatorname{rank}(F_i) = i$.

For a given sequence $a_\bullet : 1 \leq a_1 < \ldots < a_k \leq n$ of integers, consider the closed subset $D(a_\bullet)$ of X, defined by

$$D(a_\bullet) := \left\{ x \in X \,:\, \dim(E \cap F_{a_p})(x) \geq p, \ p = 1, \ldots, k \right\}.$$

As we know from the previous chapter, in the even orthogonal case there exists two families of rank n isotropic subbundles:

$$\{E \,:\, \dim(E \cap F)(x) \equiv n \,(\mathrm{mod}\ 2) \ \forall x \in X\} \quad \text{and}$$
$$\{E \,:\, \dim(E \cap F)(x) \equiv n + 1 \,(\mathrm{mod}\ 2) \ \forall x \in X\}.$$

Thus, if $a_k = n$, to get a nonempty locus, we must impose in the definition of $D(a_\bullet)$ the additional assumption that $\dim(E \cap F)(x) \equiv k \,(\mathrm{mod}\ 2)$.

A model for the $D(a_\bullet)$'s is provided by Schubert varieties in isotropic Grassmann bundles. We have three types of isotropic Grassmann bundles corresponding to the Lagrangian, odd orthogonal and even orthogonal cases respectively:

$$\pi : LG_n V \to X, \ \pi : OG_n V \to X \quad \text{and} \quad \pi : OG'_n V \to X \ (\text{resp.} \ \pi : OG''_n V \to X),$$

where $OG'_n V$ (resp. $OG''_n V$) parametrizes isotropic rank n bundles E such that $\dim(E \cap F)(x) \equiv n \, (\text{mod } 2)$ (resp. $\dim(E \cap F)(x) \equiv n+1 \, (\text{mod } 2)$). Denoting by S the tautological rank n subbundle on these Grassmann bundles, we have Schubert bundles which for $\mathcal{G} = LG_n V$ or $\mathcal{G} = OG_n V$ are defined by (we omit writing the pull-back indices)

$$\Omega(a_\bullet) := \Omega(a_\bullet, F_\bullet) = \{g \in \mathcal{G} \ : \ \dim(S \cap F_{a_i})(g) \geq i \ \forall i\}.$$

The same definition, with the additional condition $k \equiv n \, (\text{mod } 2)$, gives Schubert bundles in $\mathcal{G}' = OG'_n V$: by imposing $k \equiv n + 1 \, (\text{mod } 2)$, we get Schubert bundles in $\mathcal{G}'' = OG''_n V$. By a well-known universality property of Grassmannians there exists a morphism $s : X \to \mathcal{G}$ in the first two cases and $s = (s', s'') : X \to \mathcal{G}' \dot{\cup} \mathcal{G}''$ in the third, such that $s^* S = E$. We have set-theoretically

$$D(a_\bullet) = s^{-1}(\Omega(a_\bullet)),$$

and using the canonical scheme structure on the Schubert bundles, we define a scheme structure on $D(a_\bullet)$ by the scheme-theoretic inverse image.

The strategy used, in this chapter, to compute the fundamental classes of the $\Omega(a_\bullet)$'s and $D(a_\bullet)$'s is as follows.[1] First, we compute $[\Omega(a_\bullet)]$ using essentially two tools: a certain desingularization of $\Omega(a_\bullet)$ and the class of the (relative) diagonal

$$\Delta := \text{Im}(\delta) \subset \mathcal{G} \times_X \mathcal{G},$$

where $\delta : \mathcal{G} \to \mathcal{G} \times_X \mathcal{G}$ is the diagonal embedding. (In the even orthogonal case, $\mathcal{G}$ denotes one of the components $\mathcal{G}'$ or $\mathcal{G}''$.)

The following desingularization of $\Omega(a_\bullet)$ will be suitable to carry out the computations in the Lagrangian case (as a general rule, in this chapter, we will give a detailed account for the Lagrangian case, and just formulate analogs of the most important results for the orthogonal cases; a detailed treatment of the orthogonal cases can be found in [P-R5]).

Let $p : \mathcal{F} = \mathcal{F}(a_\bullet) \to X$ be the flag bundle parametrizing the flags

$$A_1 \subset A_2 \subset \ldots \subset A_k \subset A_{k+1}$$

where $\text{rank}(A_i) = i$ and $A_i \subset F_{a_i}$ for $i = 1, \ldots, k$; $\text{rank}(A_{k+1}) = n$ and A_{k+1} is Lagrangian. Let D denotes the tautological Lagrangian rank n vector bundle on $\mathcal{F}$. By universality of $\mathcal{G}$ there is a map $\alpha : \mathcal{F} \to \mathcal{G}$ such that $\alpha^* S = D$ (on points, α sends a flag from $\mathcal{F}$ to its last rank n member). It is easy to check that α establishes an isomorphism over the open subset of $\Omega(a_\bullet)$ parametrizing those g

---

[1] This strategy was worked out in [P4, §5], where we refer the reader for more details.

for which $\dim(S \cap F_{a_i})(g) = i$ for $i = 1, \dots, k$. The map $\alpha : \mathcal{F} \to \mathrm{Im}(\alpha) = \Omega(a_\bullet)$ is our desingularization in question. Let us consider the following diagram containing a fibre square:

$$
\begin{array}{ccc}
\Delta \subset \mathcal{G} \times_X \mathcal{G} & & \\
\qquad \uparrow {\scriptstyle 1 \times \alpha} & & \\
\mathcal{G} \times_X \mathcal{F} & \xrightarrow{\;\mathrm{pr}_2\;} & \mathcal{F} \\
\qquad {\scriptstyle \mathrm{pr}_1} \downarrow & \quad\swarrow{\scriptstyle \alpha}\quad & \downarrow {\scriptstyle p} \\
\Omega(a_\bullet) \subset \mathcal{G} & \xrightarrow[\;\pi\;]{} & X
\end{array}
$$

Note that $\sigma = \alpha \times 1$ gives a section of pr_2 and $Z = \sigma(\mathcal{F})$ maps via the proper map pr_1 birationally onto $\Omega(a_\bullet)$ (so, equivalently, $\mathrm{pr}_1|_Z : Z \to \Omega(a_\bullet)$ is a desingularization of $\Omega(a_\bullet)$). Thus knowing $[Z]$, $(\mathrm{pr}_1)_*[Z]$ gives us $[\Omega(a_\bullet)]$. But, by our construction, we have, scheme-theoretically,

$$
Z = \mathrm{Im}(\alpha \times 1) = (1 \times \alpha)^{-1}(\Delta).
$$

This is why we are interested in the class of the diagonal $[\Delta]$! Sometimes, in a situation similar to ours, there exists a vector bundle and its section such that the diagonal is the zero scheme of the section, and then one can easily apply the machinery of Chern classes (see, e.g., [K-L]). However (to the best of our knowledge and attempts), this seems not to be the case here, even for X equal to a point. If X is a point, it is pretty easy to write down a formula for $[\Delta]$. This is the content of the following

EXERCISE. Assume that X is a point. Suppose that $\{a_\alpha\}$, $\{b_\alpha\}$ are two bases of $H^*(\mathcal{G})$ that are dual w.r.t. to the Poincaré duality map $(a, b) \mapsto \int_{\mathcal{G}} a \cdot b$, that is, $\int_{\mathcal{G}} a_\alpha \cdot b_\beta = \delta_{\alpha\beta}$ (the Kronecker delta). Show that

$$
[\Delta] = \sum_\alpha a_\alpha \times b_\alpha.
$$

(Recall that since $\mathcal{G}$ has a cellular decomposition, $H^*(\mathcal{G} \times \mathcal{G}) = H^*(\mathcal{G}) \otimes_{\mathbb{Z}} H^*(\mathcal{G})$.)

From Chapter 3, we recall that the two families of Schubert classes on $\mathcal{G}$:

$$
a_\lambda := \sigma_\lambda \quad \text{and} \quad b_\lambda := \sigma_{\rho(n) \smallsetminus \lambda},
$$

both labeled by the same set of strict partitions $\lambda \subset \rho(n)$, provide a pair of bases with the above property. We remember also that

$$
\sigma_\lambda = \tilde{Q}_\lambda(S^\vee) = \pi_\lambda\big(c(S^\vee)\big)
$$

is a Giambelli-type expression in this case.

Here an interesting question appears:

Is it possible to globalize this last exercise to the relative situation?

One should be a bit careful here.

EXERCISE. Take, for example, the projectivization $\mathbb{P}(E)$ of a vector bundle E of rank $n + 1$ and consider a pair of bases

$$\left\{ c_1\left(\mathcal{O}(1)\right)^i \right\} \quad \text{and} \quad \left\{ c_1\left(\mathcal{O}(1)\right)^{n-i} \right\},$$

where both bases are labeled by $i = 0, 1, \dots, n$. These bases are Poincaré dual after restriction to each fiber. Denote by $p_i : \mathbb{P}(E) \times_X \mathbb{P}(E) \to \mathbb{P}(E)$, $i = 1, 2$, the two projections. Is

$$\sum_{i=0}^{n} p_1^* \, c_1\left(\mathcal{O}(1)\right)^i \cdot p_2^* \, c_1\left(\mathcal{O}(1)\right)^{n-i}$$

the class of the diagonal $\Delta \subset \mathbb{P}(E) \times_X \mathbb{P}(E)$?

Next, we present a criterion which says that the class of the diagonal has the "expected" form provided some "orthogonality" property w.r.t. a Gysin map is fulfilled. Let $\pi : \mathcal{G} \to X$ be a a locally trivial fiber bundle with fiber the compact (complex) manifold F. Suppose that there exist two families $\{a_\alpha\}$ and $\{b_\alpha\}$ of elements of even degree of $H^*(\mathcal{G})$ whose restrictions to the cohomology ring $H^*(F)$ of the fiber form bases of $H^*(F)$. Then, e.g., by the Leray-Hirsch theorem, $H^*(\mathcal{G} \times_X \mathcal{G})$ is a free $H^*(X)$-module with a basis $\{p_1^*(a_\alpha) \cdot p_2^*(b_\beta)\}$, where $p_i : \mathcal{G} \times_X \mathcal{G} \to \mathcal{G}$ $i = 1, 2$ are the projections onto the successive factors. In particular, the class of the diagonal $\Delta \subset \mathcal{G} \times_X \mathcal{G}$ has the form

$$[\Delta] = \sum_{\alpha, \beta} d_{\alpha\beta} \, p_1^*(a_\alpha) \cdot p_2^*(b_\beta),$$

where $d_{\alpha\beta} \in H^*(X)$.

CRITERION. With these assumptions, the following two conditions are equivalent:

(i)   For all α and β, $\pi_*(a_\alpha \cdot b_\beta) = \delta_{\alpha\beta}$.

(ii)   For all α and β, $d_{\alpha\beta} = \delta_{\alpha\beta}$.

For a proof of this result, see Appendix G.

We now sketch how one applies this criterion to our situation (where $\mathcal{G}$ is the Lagrangian Grassmann bundle and S the tautological subbundle on $\mathcal{G}$). Consider two families of elements of $H^*(\mathcal{G})$:

$$a_\lambda = \tilde{Q}_\lambda(S^\vee) \quad \text{and} \quad b_\lambda = \tilde{Q}_{\rho(n) \smallsetminus \lambda}(S^\vee),$$

where both families are labeled by all strict partitions contained in $\rho(n)$. These two families satisfy the assumptions of the above criterion. First, notice that for $|\lambda| + |\mu| < n(n + 1)/2$ we have $\pi_*(\tilde{Q}_\lambda(S^\vee) \cdot \tilde{Q}_\mu(S^\vee)) = 0$ for degree reasons. Next, observe that for $|\lambda| + |\mu| = n(n+1)/2$, $\pi_*(\tilde{Q}_\lambda(S^\vee) \cdot \tilde{Q}_\mu(S^\vee)) = 0$ unless $\mu = \rho(n) \smallsetminus \lambda$ when it is equal to 1. This is because the restrictions of $\{a_\alpha\}$ and $\{b_\beta\}$ to the cohomology ring of the fiber F are dual bases under the pairing $(x, y) \mapsto \int_F x \cdot y$

of Poincaré duality. In order to apply the above criterion, we claim that also the following property holds true:

$$\pi_* \big(\tilde{Q}_\lambda(S^\vee) \cdot \tilde{Q}_\mu(S^\vee) \big) = 0$$

for strict partitions λ, $\mu \subset \rho(n)$ such that $|\lambda| + |\mu| > n(n+1)/2$. We want to reduce this statement to a purely formal algebraic assertion. Let $X = (x_1, \ldots, x_n)$ be a sequence of variables and let $e_i(X)$ denote the ith elementary symmetric polynomial in X. Our algebraic substitute for $\tilde{Q}_\lambda$ of a vector bundle will be the $\tilde{Q}$-**polynomial** defined by the Schur Pfaffian from Chapter 3:

$$\tilde{Q}_\lambda(X) := \pi_\lambda (1 + e_1(X) + e_2(X) + e_3(X) + \ldots),$$

where λ is a (not necessary strict) partition. This is an interesting family of symmetric polynomials, not appearing explicitly in books on symmetric functions but related to the Hall-Littlewood polynomials. Let us quote some of its useful properties:

1. $\tilde{Q}_\lambda(X) \neq 0$ iff $\lambda_1 \leq n$ and $\{\tilde{Q}_\lambda(X) : \lambda_1 \leq n\}$ form a $\mathbb{Z}$-basis of the ring of symmetric polynomials in X;

2. $\tilde{Q}_{i,i}(X) = e_i(x_1^2, \ldots, x_n^2)$;

3. **Factorization property:** For two partitions $\lambda = (\lambda_1, \ldots, \lambda_k)$ and $\lambda' = (\lambda_1, \ldots, \lambda_p, i, i, \lambda_{p+1}, \ldots, \lambda_k)$, one has $\tilde{Q}_{\lambda'}(X) = \tilde{Q}_{i,i}(X) \cdot \tilde{Q}_\lambda(X)$;

4. For a more combinatorially inclined reader, perhaps the following recursive expression will be useful. For any strict partition λ,

$$\tilde{Q}_\lambda(x_1, \ldots, x_n) = \sum_{i=0}^{l(\lambda)} x_n^i \Big(\sum_{|\lambda| - |\mu| = i} \tilde{Q}_\mu(x_1, \ldots, x_{n-1}) \Big),$$

where the sum is over all (i.e. not necessarily strict) partitions $\mu \subset \lambda$ such that $D_\lambda \smallsetminus D_\mu$ has at most one box in every row.

Moreover, let us record the following, more difficult, multiplication property. We will refer to it as Pieri-type formula for $\tilde{Q}$-polynomials.

5. For a strict partition $\lambda = (\lambda_1 > \ldots > \lambda_k > 0)$,

$$\tilde{Q}_r(X) \cdot \tilde{Q}_\lambda(X) = \sum 2^{m(\lambda, \mu)} \tilde{Q}_\mu(X),$$

where the sum is over partitions $\mu \supset \lambda$, $|\mu| = |\lambda| + r$, $D_\mu \smallsetminus D_\lambda$ is a horizontal strip and $m(\lambda, \mu)$ is the number of its connected components not meeting the first column.

In other words, this multiplication property is the same as for Schur Q-polynomials (see Chapter 3) but we allow non-strict partitions to appear in the product.

EXAMPLE. Writing temporarily $\widetilde{Q}_\lambda = \widetilde{Q}_\lambda(X)$, the product

$$\widetilde{Q}_5 \cdot \widetilde{Q}_{6,3,2}$$

contains, besides of the following summands labeled by strict partitions:

$$2\widetilde{Q}_{11,3,2} + 4\widetilde{Q}_{10,4,2} + 2\widetilde{Q}_{10,3,2,1} + 4\widetilde{Q}_{9,5,2} + 4\widetilde{Q}_{9,4,3} + 4\widetilde{Q}_{9,4,2,1}$$
$$+ 2\widetilde{Q}_{8,6,2} + 4\widetilde{Q}_{8,5,3} + 4\widetilde{Q}_{8,5,2,1} + 4\widetilde{Q}_{8,4,3,1}$$
$$+ 2\widetilde{Q}_{7,6,3} + 2\widetilde{Q}_{7,6,2,1} + 4\widetilde{Q}_{7,5,3,1} + 2\widetilde{Q}_{7,4,3,2} + \widetilde{Q}_{6,5,3,2}$$

also the following summands labeled by non-strict partitions:

$$+ 2\widetilde{Q}_{6,6,3,1} + 2\widetilde{Q}_{6,6,2,2} + 4\widetilde{Q}_{10,3,3} + 4\widetilde{Q}_{9,3,3,1} + 2\widetilde{Q}_{8,3,3,2}$$
$$+ 2\widetilde{Q}_{9,3,2,2} + 4\widetilde{Q}_{8,4,2,2} + 2\widetilde{Q}_{7,5,2,2}.$$

For the proofs of properties 1–5 and for more on $\widetilde{Q}$-polynomials, see [P-R5].

Our algebraic substitute for π_* will be a certain symplectic divided difference operator. Before we introduce it, we need some notation associated with the **hyperoctahedral group**, i.e. the Weyl group of the symplectic group. Recall that the symmetric group S_n (of all bijections of $\{1, \dots, n\}$) is generated by the simple transpositions s_i that interchange i and $i+1$ for $i = 1, \dots, n-1$. The hyperoctahedral group $W = W_n$ is an extension of S_n by an element s_0 such that the following "braid relations" hold

$$s_0^2 = 1, \quad s_0 \cdot s_1 \cdot s_0 \cdot s_1 = s_1 \cdot s_0 \cdot s_1 \cdot s_0, \quad s_0 \cdot s_i = s_i \cdot s_0 \quad \text{for} \quad i \geq 2.$$

More precisely, $W = S_n \ltimes \mathbb{Z}_2^n$ where S_n acts on $\mathbb{Z}_2^n$ by permutations. Writing a typical element of W as $w = (\sigma, \tau)$ with $\sigma \in S_n$ and $\tau \in \mathbb{Z}_2^n$, we have for $w' = (\sigma', \tau')$

$$w \cdot w' = (\sigma\sigma', \delta),$$

where $\delta_i = \tau_{\sigma'(i)} \cdot \tau_i'$. To represent elements of W we use the standard "barred-permutation" notation, writing them as permutations with bars in those places (numbered with "i") where $\tau_i = -1$. With this notation we have $s_0 = [\bar{1}, 2, \dots, n]$, and the right multiplication by s_0 is given by

$$[w_1, w_2, \dots, w_n] \cdot s_0 = [\bar{w}_1, w_2, \dots, w_n];$$

also $s_i = [1, \dots, i-1, i+1, i, i+2, \dots, n]$ with

$$[w_1, w_2, \dots, w_n] \cdot s_i = [w_1, \dots, w_{i-1}, w_{i+1}, w_i, w_{i+2}, \dots, w_n].$$

In this notation, the longest element w_0 of W becomes $w_0 = [\bar{1}, \dots, \bar{n}]$ and the formula for the length $l(w)$ of $w = (\sigma, \tau) \in W$ reads:

$$l(w) = \sum_{i=1}^n a_i + \sum_{\{j : \tau_j = -1\}} (2b_j + 1),$$

where $a_i = \text{card}\{p : p > i \ \& \ \sigma_p < \sigma_i\}$ and $b_j = \text{card}\{p : p < j \ \& \ \sigma_p < \sigma_j\}$.[2]

Let $X = (x_1, \ldots, x_n)$ be a sequence of variables. The group algebra of W acts on the ring $\mathbb{Z}[X]$ of polynomials in X: s_i interchanges x_i and x_{i+1}, $1 \le i \le n-1$, and s_0 replaces x_1 by $-x_1$ (all other variables remaining unchanged).

In the present chapter, we shall make an extensive use of the algebra of divided differences, with generators $\partial_0, \partial_1, \ldots, \partial_{n-1}$, acting on the ring $\mathbb{Z}[X]$ according to

$$\partial_i(P) = \bigl(P - s_i(P)\bigr)/(x_i - x_{i+1}), \quad i \ge 1,$$

$$\text{and} \quad \partial_0(P) = \bigl(P - s_0(P)\bigr)/(-2x_1).$$

These generators satisfy the braid relations for the group W. Consequently, for every $w \in W$, there exists a well-defined divided difference $\partial_w = \partial_{i_1} \circ \ldots \circ \partial_{i_l}$ which can be obtained from any reduced decomposition $s_{i_1} \cdot \ldots \cdot s_{i_l}$ of w (see [B-G-G] and [D1,2]).[3]

In particular, consider the following divided difference operator

$$\nabla = \partial_{[\bar{n}, \overline{n-1}, \ldots, \bar{1}]} : \mathbb{Z}[X] \to \mathbb{Z}[X],$$

that is,

$$\nabla = \partial_0(\partial_1 \partial_0) \ldots (\partial_{n-1} \ldots \partial_1 \partial_0) = (\partial_0 \partial_1 \ldots \partial_{n-1}) \ldots (\partial_0 \partial_1)\partial_0.$$

EXERCISE. Show that $\partial_{[n,\ldots,1]}$ coincides with the Jacobi symmetrizer ∂

$$P \mapsto \sum_{w \in S_n} w[P \big/ \prod_{1 \le i < j \le n} (x_i - x_j)].$$

Similarly, prove that ∂_{w_0} coincides with

$$P \mapsto (-1)^{n(n+1)/2} \sum_{w \in W} w[P \big/ 2^n x_1 \cdot \ldots \cdot x_n \prod_{1 \le i < j \le n} (x_i^2 - x_j^2)].$$

Using $\partial_{w_0} = \nabla \circ \partial_{[n,\ldots,1]}$, show that for every symmetric polynomial P, the operator ∇ acts as

$$P \mapsto (-1)^{n(n+1)/2} \sum_{w \in W/S_n} w[P \big/ \prod_{1 \le i \le j \le n} (x_i + x_j)].$$

(Consult also [D1] and [P-R5].)

Clearly, the operator ∇ decreases degree by $n(n+1)/2$ and for any polynomial P symmetric in $x_1^2, \ldots, x_n^2$ and any polynomial Q, we have

$$\nabla(P \cdot Q) = P \cdot \nabla(Q).$$

---

[2]The isomorphism with the subgroup W of S_{2n} used in Chapter 6 interchanges s_i and s_{n-i} for all i. Given $v \in W \subset S_{2n}$, the corresponding element $[w_1, \ldots, w_n]$ is defined by the rule: $w_i = n + 1 - v(n + 1 - i)$ if $v(n + 1 - i) \le n$, and $w_i = \overline{v(n + 1 - i) - n}$ otherwise.

[3]To make the divided difference operators agree with those of Chapter 6, with $\partial_i \leftrightarrow \partial_{n-i}$, interchange x_i with $-x_{n+1-i}$ for all i.

Moreover, ∇ has the following useful geometric interpretation: it induces π_*, where $\pi : LG_nV \to X$. That is, for every symmetric polynomial P,

$$\pi_*\big(P(r_1,\dots,r_n)\big) = (\nabla P)(r_1,\dots,r_n),$$

where $r_1,\dots,r_n$ are the Chern roots of the tautological bundle S on LG_nV. Perhaps the simplest way to see this is to consider the bundle of complete isotropic flags

$$\tau : LFl(V) \xrightarrow{\ \omega\ } \mathcal{G} = LG_nV \xrightarrow{\ \pi\ } X$$

and invoke that ω_* is induced by the Jacobi symmetrizer, and analogously τ_* is induced by the symmetrizer ∂_{w_0} (see Chapter 4 and [Br]).

Let, from now on, $\widetilde{Q}_\lambda = \widetilde{Q}_\lambda(-x_1,\dots,-x_n)$. We claim that for strict partitions $\lambda,\ \mu \subset \rho(n)$ such that $|\lambda| + |\mu| > n(n+1)/2$, one has

(7.1) $$\nabla(\widetilde{Q}_\lambda \cdot \widetilde{Q}_\mu) = 0.$$

The proof is by double induction on $l(\lambda)$ and the shortest part of λ, that is $\lambda_{l(\lambda)}$.

EXERCISE. Prove that the identity (7.1) holds true if λ has one part only. (Hint: Use the Pieri-type formula 5 and the factorization property 3.)

To perform the inductive step, we set $l = l(\lambda)$ and write

$$\lambda' = (\lambda_1,\dots,\lambda_{l-1}) \quad \text{and} \quad r = \lambda_l.$$

Now we use twice the Pieri-type formula 5 and write

(7.2)
$$\widetilde{Q}_\lambda \cdot \widetilde{Q}_\mu = \widetilde{Q}_{\lambda'} \cdot \widetilde{Q}_r \cdot \widetilde{Q}_\mu - \left(\sum_\alpha 2^{m(\lambda',\alpha)}\widetilde{Q}_\alpha\right) \cdot \widetilde{Q}_\mu$$

$$= \widetilde{Q}_{\lambda'} \cdot \left(\sum_\beta 2^{m(\mu,\beta)}\widetilde{Q}_\beta\right) - \left(\sum_\alpha 2^{m(\lambda',\alpha)}\widetilde{Q}_\alpha\right) \cdot \widetilde{Q}_\mu .$$

In the last line of (7.2), besides other restrictions imposed on the α's and the β's by the Pieri-type formula, all α are different from λ, and we can apply the induction assumption to them. Recall that the α's and β's can have pairs of repeated parts (in fact each such a pair can occur only once). If this happens, we use the factorization property 3 w.r.t. all such pairs. Using the induction assumption, or because of degree reasons, we then see that after applying ∇ to the last line of (7.2), the following holds true. There survive in the first sum only the products

$$2^{m(\mu,\beta)}\widetilde{Q}_{\lambda'} \cdot \widetilde{Q}_\beta,$$

where the partition obtained by removing all repeated parts from β is equal to $\rho(n) \smallsetminus \lambda'$. Similarly, there survive in the second sum only the products

$$-2^{m(\lambda',\alpha)}\widetilde{Q}_\alpha \cdot \widetilde{Q}_\mu,$$

where the partition obtained by removing all repeated parts from α is equal to $\rho(n) \smallsetminus \mu$. It is enough to give a bijection between the set $\mathcal{B}$ of such β's and the set $\mathcal{A}$ of such α's. Such a bijection is defined as follows. Let $\beta \in \mathcal{B}$. Remove from β one part from each pair of repeated parts. Take the complement in $\rho(n)$ of the so-obtained partition and insert into this complement the previously removed parts in such a way that we obtain a partition. This partition belongs to $\mathcal{A}$ and is the image of β via our bijection.

By reversing the roles of $\mathcal{B}$ and $\mathcal{A}$, we get the inverse map.

EXAMPLE OF THE CORRESPONDENCE $\mathcal{A} \leftrightarrow \mathcal{B}$.

$n = 10$, $\beta = (10, 9, 7, 4, 4, 2, 1)$, $\mu = (10, 8, 6, 4, 2, 1)$, $r = 6$:

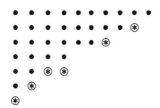

the complement in $\rho(10)$ of $(10, 8, 6, 2, 1)$ is $(9, 7, 5, 4, 3)$:

So $\alpha = (9, 7, 5, 4, 4, 3)$ (and $\lambda' = (8, 6, 5, 4, 3)$):

It is transparent from this example that our bijection preserves the number of the connected components of the r-strip not meeting the first column. For more details, see [P-R5].

This orthogonality property gives, via our criterion, that the class of the diagonal in $\mathcal{G} \times_X \mathcal{G}$ equals

$$[\Delta] = \sum \tilde{Q}_\lambda(S_1^\vee) \cdot \tilde{Q}_{\rho(n) \smallsetminus \lambda}(S_2^\vee),$$

where the sum is over all strict partitions $\lambda \subset \rho(n)$ and S_i are the tautological bundles on the corresponding factors. Consequently, in $H^*(\mathcal{G} \times_X \mathcal{F})$,

$$[Z] = \sum \tilde{Q}_\lambda(S^\vee) \cdot \tilde{Q}_{\rho(n) \smallsetminus \lambda}(D^\vee),$$

where the sum is over all strict partitions $\lambda \subset \rho(n)$ (and we omit the pull-back indices in both formulas).

To complete the calculation of the class of $\Omega(a_\bullet)$, we need an explicit expression for $(\mathrm{pr}_1)_* \tilde{Q}_\lambda(D^\vee)$, or equivalently – using base change – for $p_* \tilde{Q}_\lambda(D^\vee)$.

We give first an explicit description for $a_\bullet = (n - k + 1, n - k + 2, \ldots, n)$. For a strict partition $\lambda \subset \rho(n)$, if $(n, n-1, \ldots, k+1) \not\subset \lambda$ then $p_* \tilde{Q}_\lambda(D^\vee) = 0$. Otherwise, if $\lambda = (n, n-1, \ldots, k+1, \mu_1, \ldots, \mu_l)$, then with $\mu = (\mu_1, \ldots, \mu_l)$,

$$(7.3) \qquad\qquad p_* \tilde{Q}_\lambda(D^\vee) = \tilde{Q}_\mu(F^\vee).$$

The reader can find in [P-R5] two proofs of (7.3). The first is a Schubert-calculus-type proof using linear algebra arguments. The second is similar to the above proof of the orthogonality property; it involves the symplectic divided difference operator $\partial_{w^{(k)}}$, where $w^{(k)}$ is the barred permutation

$$w^{(k)} = [\bar{n}, \overline{n-1}, \ldots, \overline{k+1}, 1, 2, \ldots, k].$$

The operator $\partial_{w^{(k)}}$ induces p_* in the following sense. For any symmetric polynomial P in n variables,

$$p_* P(d_1, \ldots, d_n) = \left(\partial_{w^{(k)}} P\right)(r_1, \ldots, r_n),$$

where $d_1, \ldots, d_n$ is the sequence of Chern roots of D, and $r_1, \ldots, r_n$ are the Chern roots of F. The claimed formula (7.3) then follows from a purely formal algebraic result that if $(n, n-1, \ldots, k+1) \not\subset \lambda$, then $\partial_{w^{(k)}} \tilde{Q}_\lambda = 0$; and if $\lambda = (n, n-1, \ldots, k+1, \mu_1, \ldots, \mu_l)$, then

$$(7.4) \qquad\qquad \partial_{w^{(k)}} \tilde{Q}_\lambda = \tilde{Q}_\mu.$$

EXERCISE. Prove the identity (7.4) for $k = n - 1$; that is,

$$\partial_{n-1} \ldots \partial_1 \partial_0 (\tilde{Q}_{n,\mu}) = \tilde{Q}_\mu.$$

(Hint: Use the Leibniz-type rule for divided differences.)

In the case $a_\bullet = (n - k + 1, \ldots, n)$, after applying the formula for $(\mathrm{pr}_1)_*$ to the above expression for $[Z]$, we get

$$[\Omega(n - k + 1, \ldots, n)] = \sum_\lambda \tilde{Q}_\lambda(F^\vee) \cdot \tilde{Q}_{\rho(k) \smallsetminus \lambda}(S^\vee),$$

where the sum is over strict partitions $\lambda \subset \rho(k)$. Set $\tilde{P}_\lambda(E) := 2^{-l(\lambda)} \tilde{Q}_\lambda(E)$ for a bundle E. Analogous expressions for the orthogonal cases are, in the odd orthogonal case,

$$[\Omega(n - k + 1, \ldots, n)] = \sum_\lambda \tilde{P}_\lambda(F^\vee) \cdot \tilde{P}_{\rho(k) \smallsetminus \lambda}(S^\vee),$$

and in the even orthogonal case,

$$[\Omega(n - k + 1, \ldots, n)] = \sum_\lambda \tilde{P}_\lambda(F^\vee) \cdot \tilde{P}_{\rho(k-1) \smallsetminus \lambda}(S^\vee),$$

where for $k \equiv n \pmod 2$ the equation holds in $H^*(OG'_n E)$, and for $k \equiv n + 1 \pmod 2$ – in $H^*(OG''_n E)$. In the former sum, the summation is over strict partitions $\lambda \subset \rho(k)$ and in the latter – over strict partitions $\lambda \subset \rho(k-1)$.

Finally, we are in position to give the following answer to the problem of Harris.

Let D_k be the scheme $D(n - k + 1, n - k + 2, \ldots, n)$ supported by the locus $\{x \in X : \dim(E \cap F)(x) \geq k\}$. (Recall that in the even orthogonal case we assume $k \equiv \dim(E \cap F)(x) \pmod 2$ for $x \in X$.) Then using the section $s : X \to \mathcal{G}$ such that $s^* S = E$ and the lemma from Appendix A, we pull back the formulas from $H^*(\mathcal{G})$ to $H^*(X)$. We get:

THEOREM. *The loci D_k are represented by the following polynomials:*
(i) in the Lagrangian case,

$$(7.5) \qquad [D_k] = \sum_{\lambda \subset \rho(k)} \tilde{Q}_\lambda(E^\vee) \cdot \tilde{Q}_{\rho(k) \smallsetminus \lambda}(F^\vee);$$

(ii) in the odd orthogonal case,

$$(7.6) \qquad [D_k] = \sum_{\lambda \subset \rho(k)} \tilde{P}_\lambda(E^\vee) \cdot \tilde{P}_{\rho(k) \smallsetminus \lambda}(F^\vee);$$

(iii) in the even orthogonal case,

$$(7.7) \qquad [D_k] = \sum_{\lambda \subset \rho(k-1)} \tilde{P}_\lambda(E^\vee) \cdot \tilde{P}_{\rho(k-1) \smallsetminus \lambda}(F^\vee),$$

all sums being over strict partitions λ.

For example, the expressions answering the original Harris' problem for successive k are:

k=1   1;

k=2 $\tilde{P}_1(E^\vee) + \tilde{P}_1(F^\vee)$;

k=3 $\tilde{P}_{21}(E^\vee) + \tilde{P}_2(E^\vee) \cdot \tilde{P}_1(F^\vee) + \tilde{P}_1(E^\vee) \cdot \tilde{P}_2(F^\vee) + \tilde{P}_{21}(F^\vee)$;

k=4 $\tilde{P}_{321}(E^\vee) + \tilde{P}_{32}(E^\vee) \cdot \tilde{P}_1(F^\vee) + \tilde{P}_{31}(E^\vee) \cdot \tilde{P}_2(F^\vee) + \tilde{P}_{21}(E^\vee) \cdot \tilde{P}_3(F^\vee) +$
$\tilde{P}_3(E^\vee) \cdot \tilde{P}_{21}(F^\vee) + \tilde{P}_2(E^\vee) \cdot \tilde{P}_{31}(F^\vee) + \tilde{P}_1(E^\vee) \cdot \tilde{P}_{32}(F^\vee) + \tilde{P}_{321}(F^\vee).$

These expressions will be used in the next chapter to compute the cohomology classes of Brill-Noether loci in Prym varieties.

The reader is invited to try to compute, along the above lines, the expressions for $[\Omega(a_\bullet)]$ for some other $a_\bullet$'s. All needed tools have been given here.

For example, in the case of a single Schubert condition $a = n + 1 - i$ for the Lagrangian Grassmannian $\mathcal{G} = LG_n V$, one gets

$$[\Omega(a)] = \sum_{p=0}^{i} c_p(S^\vee) \cdot s_{i-p}(F_a^\vee),$$

where F_a is an isotropic subbundle of V of rank a, involved in the definition of $\Omega(a)$. This can be done by pushing forward the class

$$[Z] = \sum_{\text{strict } \lambda \subset \rho(n)} \tilde{Q}_\lambda(D^\vee) \cdot \tilde{Q}_{\rho(n) \smallsetminus \lambda}(S^\vee)$$

in the composite

$$\mathcal{F} = LG_{n-1}(C^\perp/C) \to \mathbb{P}(F_a) \to \mathcal{G},$$

where C is the tautological line bundle on $\mathbb{P}(F_a)$.

For $G = OG_n V$, where $\dim(V) = 2n + 1$, one gets for $a = n + 1 - i$,

$$[\Omega(a)] = 1/2 \cdot \sum_{p=0}^{i} c_p(S^\vee)\, s_{i-p}(F_a^\vee).$$

The class of $\Omega(a)$ in $H^*(OG'_n V)$ for odd n, or in $H^*(OG''_n V)$ for even n, is given by the following expression where $i = n - a$:

$$[\Omega(a)] = 1/2 \cdot \sum_{p=0}^{i} (c_p(S^\vee) + c_p(F))\, s_{i-p}(F_a^\vee).$$

(To prove this, one must use the result of Edidin-Graham explained in Chapter 6.)

A globalization of, for instance, the first of these formulas, using the setup from the beginning of this chapter, reads: the degeneracy locus

$$\{x \in X : \dim(E \cap F_{n+1-i})(x) \geq 1\}$$

is represented by the polynomial

$$\sum_{p=0}^{i} c_p(E^\vee) \cdot s_{i-p}(F_{n+1-i}^\vee).$$

For example, the expressions giving the classes for successive i are:

$i=1$ $c_1(E^\vee) + s_1(F^\vee)$;

$i=2$ $c_2(E^\vee) + c_1(E^\vee)s_1(F_{n-1}^\vee) + s_2(F_{n-1}^\vee)$;

$i=3$ $c_3(E^\vee) + c_2(E^\vee)s_1(F_{n-2}^\vee) + c_1(E^\vee)s_2(F_{n-2}^\vee) + s_3(F_{n-2}^\vee)$.

The reader can find in [P-R5] more examples of such explicit computations.

The method sketched in this chapter offers also the following approach to **symplectic Schubert polynomials**. (Recall that other approaches were given by Billey-Haiman, Fomin-Kirillov and Fulton - see Section 9.5). Let $(x_1, x_2, \dots)$ be a sequence of variables. Let w_0 be the longest element in the Weyl group W of type C_n. Define

$$\mathcal{C}_{w_0} = \mathcal{C}_{w_0}(x_1, \dots, x_n) := (-1)^{n(n-1)/2}\, x^{\rho(n-1)}\, \tilde{Q}_{\rho(n)}(x_1, \dots, x_n),$$

and for an arbitrary $w \in W$,

$$C_w = C_w(x_1, \ldots, x_n) := \partial'_{w^{-1}w_0}(C_{w_0}).$$

Above, by ∂'_w ($w \in W$) we understand the composition of the divided difference operators $\partial'_i := -\partial_i$ for $i = 1, \ldots, n-1$ and $\partial'_0 := \partial_0$ – to make the signs more convenient.

These polynomials satisfy the following properties.

1. (Stability) Suppose that $m > n$. Let $\iota : W_n \hookrightarrow W_m$ be the embedding via the first n components. Then, for any $w \in W_n$, the following equality holds:

$$C_{\iota(w)}(x_1, \ldots, x_m)|_{x_{n+1}=\ldots=x_m=0} = C_w(x_1, \ldots, x_n).$$

2. (The maximal Grassmannian property) Let $\lambda = (\lambda_1 > \ldots > \lambda_k > 0)$ be a strict partition contained in $\rho(n)$. Set

$$w_\lambda := [\overline{\lambda_1}, \ldots, \overline{\lambda_k}, \mu_1 < \mu_2 < \ldots < \mu_{n-k}],$$

where $\{\lambda_1, \ldots, \lambda_k, \mu_1, \ldots, \mu_{n-k}\} = \{1, 2, \ldots, n\}$. Then

$$C_{w_\lambda}(x_1, \ldots, x_n) = \tilde{Q}_\lambda(x_1, \ldots, x_n).$$

For $n = 2$ these polynomials are:

$$C_{[\overline{1},\overline{2}]} = -x_1^3 x_2 - x_1^2 x_2^2$$

$$C_{[1,\overline{2}]} = -x_1^2 x_2 \qquad C_{[\overline{2},\overline{1}]} = x_1^2 x_2 + x_1 x_2^2$$

$$C_{[\overline{2},1]} = x_1 x_2 \qquad C_{[2,\overline{1}]} = x_2^2$$

$$C_{[2,1]} = x_2 \qquad C_{[\overline{1},2]} = x_1 + x_2$$

$$C_{[1,2]} = 1.$$

These symplectic Schubert polynomials should be useful for geometers because of the Grassmannian property 2. Also, they contain fewer monomials than the other known families, which make them handier for numerical computations. We refer the reader to [L-P] for proofs and more details.

REMARK. Set $\tilde{P}_\lambda(x_1, \ldots, x_n) := 2^{-l(\lambda)} \tilde{Q}_\lambda(x_1, \ldots, x_n)$. If one makes the following changes in the theorems of Chapter 6:

in Theorem(C_n) one replaces $\Delta_{\rho(n)}(c)$ by

$$\sum_\lambda \tilde{Q}_\lambda(x_1, \ldots, x_n) \tilde{Q}_{\rho(n)\smallsetminus\lambda}(y_1, \ldots, y_n);$$

in Theorem(B_n) one replaces $\Delta_{\rho(n)}(c)$ by

$$\sum_\lambda \tilde{P}_\lambda(x_1, \ldots, x_n) \tilde{P}_{\rho(n)\smallsetminus\lambda}(y_1, \ldots, y_n);$$

in Theorem(D_n) one replaces $\Delta_{\rho(n-1)}(c)$ by

$$\sum_\lambda \tilde{P}_\lambda(x_1, \ldots, x_n) \tilde{P}_{\rho(n-1)\smallsetminus\lambda}(y_1, \ldots, y_n),$$

where the first two sums are over strict partitions $\lambda \subset \rho(n)$ and the last summation is over strict partitons $\lambda \subset \rho(n-1)$, then these theorems remain true. For more on this, consult [L-P].

CHAPTER VIII

THE CLASSES OF BRILL-NOETHER
LOCI IN PRYM VARIETIES

In this chapter we apply a formula from Chapter 7 to Brill-Noether loci in Prym varieties. Let us start with a brief definition of the Prym variety associated with a double unramified cover $\pi : \widetilde{C} \to C$ of a smooth algebraic curve C over $\mathbb{C}$ (in fact, most of the results holds over an algebraically closed field of characteristic different from 2). With such a double cover there is attached a norm map. The assignment $\mathrm{Nm} : \mathrm{Div}(\widetilde{C}) \to \mathrm{Div}(C)$ such that $\sum m_i P_i \mapsto \sum m_i \pi(P_i)$ has the property $\mathrm{Nm}(\mathrm{Div}(f)) = \mathrm{Div}(\mathrm{Nm}(f))$, and thus induces a map of the corresponding Jacobians $\mathcal{J}(\widetilde{C}) \to \mathcal{J}(C)$ called the **norm** of π and also denoted by Nm. Equivalently, on the level of Picard groups the norm of the (class) of a line bundle L on $\widetilde{C}$ is the class of $\wedge^2(\pi_* L)$ (and this definition works also for general unramified double covers).

Let $\pi^* : \mathcal{J}(C) \to \mathcal{J}(\widetilde{C})$ be the map induced by π and let $\tau : \widetilde{C} \to \widetilde{C}$ denote the involution exchanging the sheets of $\widetilde{C}$ over C. Denote by the same letter the induced action of τ on $\mathcal{J}(\widetilde{C})$. For any divisor D on $\widetilde{C}$, we have $\pi^{-1}(\pi D) = D + \tau D$, so $\pi^*(\mathrm{Nm}\, x) = x + \tau x$ for $x \in \mathcal{J}(\widetilde{C})$. It follows easily that τ acts on $\pi^* \mathcal{J}(C)$ by the identity and τ acts on $\mathrm{Ker}(\mathrm{Nm})$ by the minus identity.

EXERCISE. Prove these last two assertions.

We define the **Prym variety** P (associated with π) as $(\mathrm{Ker}(\mathrm{Nm}))^0$ – the connected component of $\mathrm{Ker}(\mathrm{Nm})$ containing the origin. So equivalently,

$$P = \mathrm{Ker}\,(\mathrm{id}_{\mathcal{J}(\widetilde{C})} + \tau)^0 = \mathrm{Im}(\mathrm{id}_{\mathcal{J}(\widetilde{C})} - \tau);$$

thus intuitively P is the antiinvariant or "odd" part of $\mathcal{J}(\widetilde{C})$. If the genus of C is g, then (by the Riemann-Hurwitz formula) the genus of $\widetilde{C}$ is $2g-1$. Consequently, P is an abelian variety of dimension $g-1$. (For those who prefer the analytic approach, by identyfying $\mathcal{J}(\widetilde{C})$ with $H^0(\widetilde{C}, \Omega_{\widetilde{C}})^\vee / H_1(\widetilde{C}, \mathbb{Z})$, τ acts on both the groups and the Prym variety is gotten by integration of the basic antiinvariant forms over the basic antiinvariant cycles:

$$P = \left(H^0(\widetilde{C}, \Omega_{\widetilde{C}})^-\right)^\vee / H_1(\widetilde{C}, \mathbb{Z})^-.)$$

For more about Prym varieties, see [A-C-G-H, App.C] and [Mu2].

To define Brill-Noether loci for Prym varieties, it is convenient to work with Picard varieties parametrizing divisor classes (or equivalently, classes of line bundles) of degree $2g - 2$. Let $\omega_C \in \text{Pic}^{2g-2}(C)$ be the canonical class. Then, denoting by $\text{Nm} : \text{Pic}^{2g-2}(\widetilde{C}) \to \text{Pic}^{2g-2}(C)$ the norm, we have

$$\text{Nm}^{-1}(\omega_C) = P^+ \,\dot\cup\, P^-,$$

where $P^+ = \{L : h^0(\widetilde{C}, L) \text{ is even}\}$ and $P^- = \{L : h^0(\widetilde{C}, L) \text{ is odd}\}$. The varieties P^+, P^- are translates of P, and thus they are irreducible of dimension $g - 1$. In this chapter, $P^\pm$ will denote either of these varieties (and not their union).

The following definition of **Brill-Noether loci** is due to Welters (see [W]). For a given integer $r \geq -1$, we set

$$V^r = V^r(\pi) := \{ L \in P^\pm \,:\, h^0(\widetilde{C}, L) \geq r + 1 \ \& \ h^0(\widetilde{C}, L) \equiv r + 1 (\text{mod } 2)\}.$$

(Note that the last condition is imposed to get a connected set.) We have $V^{-1} = P^+$, $V^0 = P^-$ and $V^r \subset P^+$ if and only if r is odd.

The following main result about the structure of V^r was established by Welters.

THEOREM. *Suppose that $\mathcal{L}$ is a Poincaré line bundle on $\text{Pic}^{2g-2}(\widetilde{C}) \times \widetilde{C}$ and let*

$$\nu : \text{Pic}^{2g-2}(\widetilde{C}) \times \widetilde{C} \to \text{Pic}^{2g-2}(\widetilde{C})$$

be the projection. Equip V^r with the scheme structure defined by the $(r+1)^{st}$ Fitting ideal of $\mathcal{R}^1\nu_*\mathcal{L}\,|_{P^\pm}$. Then for a general curve C and any irreducible double cover of it, V^r, if nonempty, has pure codimension $r(r + 1)/2$ and $V^r \smallsetminus V^{r+2}$ is smooth.

For a proof of this theorem, see [W]. Later on, we will define a "more reduced" scheme structure on V^r. This scheme structure will be a natural by-product of another presentation of V^r that we describe now.

We start with a representative $\mathcal{L}$ of the Poincaré bundle on $\text{Pic}^{2g-2}(\widetilde{C}) \times \widetilde{C}$ such that for some $c \in \widetilde{C}$,

$$\mathcal{L}\,|_{\text{Pic}^{2g-2}(\widetilde{C}) \times \{c\}} \in \text{Pic}^0(\text{Pic}^{2g-2}(\widetilde{C})),$$

and for the norm Nm associated with $1 \times \pi$ one has $\text{Nm}\,\mathcal{L}\,|_{P^\pm \times C} \cong q^*\Omega_C$, where $q : \text{Pic}^{2g-2}(\widetilde{C}) \times C \to C$ is the projection. As Fulton points out, such a representative can be constructed in the following way. If $\mathcal{L}_0$ is a representative of the Poincaré bundle such that

$$\mathcal{L}_0\,|_{\text{Pic}^{2g-2}(\widetilde{C}) \times \{c\}} \in \text{Pic}^0(\text{Pic}^{2g-2}(\widetilde{C}))$$

and for any $L \in P^\pm$, $\text{Nm}\,\mathcal{L}_0\,|_{\{L\} \times C} \cong \Omega_C$, then there exists $M \in \text{Pic}^0(\text{Pic}^{2g-2}(\widetilde{C}))$ such that

$$\text{Nm}\,\mathcal{L}_0\,|_{P^\pm \times C} \cong p^*M \otimes q^*\Omega_C\,|_{P^\pm \times C},$$

where $p : \text{Pic}^{2g-2}\widetilde{C} \times C \to \text{Pic}^{2g-2}\widetilde{C}$ is the projection. Since $\text{Pic}^{2g-2}(\widetilde{C})$ is isomorphic to an abelian variety, there exists $L \in \text{Pic}^0(\text{Pic}^{2g-2}(\widetilde{C}))$ such that $M = L^{\otimes 2}$. Then $\mathcal{L} = \mathcal{L}_0 \otimes (p')^*(L^\vee)$, where $p' : \text{Pic}^{2g-2}(\widetilde{C}) \times \widetilde{C} \to \text{Pic}^{2g-2}(\widetilde{C})$ is the projection, does the job.

Set $\mathcal{E} = (1 \times \pi)_* \mathcal{L}$. We claim that $\mathcal{E}|_{P^\pm \times C}$ is equipped with a certain natural quadratic form Q with values in $q^* \Omega_C|_{P^\pm \times C}$. Fix $L \in P^\pm$. We explain what is the value of Q on $\{L\} \times C$. Since $\mathrm{Nm}\, L \cong \Omega_C$, we have

$$L \otimes \tau^* L = \pi^* \mathrm{Nm}\, L \cong \pi^* \Omega_C = \Omega_{\widetilde{C}}.$$

Let us denote the composition of these isomorphisms by α. Set $E = \pi_* L$. Then our goal is to define a quadratic form $E \otimes E \to \Omega_C$. Consider any open subset $U \subset C$ and let $\widetilde{U} = \pi^{-1}(U)$. We define first the bilinear form

$$\Gamma(\widetilde{U}, L) \otimes \Gamma(\widetilde{U}, L) \to \Gamma(\widetilde{U}, \Omega_{\widetilde{C}})$$

by $s \otimes t \mapsto \alpha(s \otimes \tau^* t) =:<s, t>$. We have $\tau^*(<s, t>) =<t, s>$. Since τ^* acts on $\Gamma(\widetilde{U}, \Omega_{\widetilde{C}})$ with $(+1)$-eigenspace $\Gamma(U, \Omega_C)$, restricting the above bilinear form to $S^2 \Gamma(\widetilde{U}, L)$, we get the required nondegenerate quadratic form

$$\Gamma(\widetilde{U}, L) \otimes \Gamma(\widetilde{U}, L) \to \Gamma(U, \Omega_C).$$

For more on this, see [Mu2].

Let now D be a sufficiently positive divisor $\sum P_i$ on C, where the P_i's are distinct. Recalling that $p : \mathrm{Pic}^{2g-2}(\widetilde{C}) \times C \to C$ is the projection, we set

$$V := p_*(\mathcal{E}(D)/\mathcal{E}(-D))|_{P^\pm}, \quad W := p_* \mathcal{E}(D)|_{P^\pm}, \quad U := p_*(\mathcal{E}/\mathcal{E}(-D))|_{P^\pm},$$

where $\mathcal{E}(\pm D) = \mathcal{E} \otimes q^* \mathcal{O}_C(\pm D)$. We use now Mumford's construction from Appendix H. At a fixed $L \in P^\pm$, it assigns to the above quadratic form $E \otimes E \to \Omega_C$ a nondegenerate quadratic form on $\Gamma(C, E(D)/E(-D))$ and two maximal isotropic subspaces $\Gamma(C, E(D))$ and $\Gamma(C, E/E(-D))$ of it such that $\Gamma(C, E)$ is equal to their intersection. Thanks to our choice of $\mathcal{L}$, this construction globalizes, and we get a nondegenerate quadratic form $\Phi : V \otimes V \to \mathcal{O}_{P^\pm}$ such that U and W are maximal isotropic subbundles of V, and the Brill-Noether locus $V^r \subset P^\pm$ has a presentation

$$V^r = \{\, L \in P^\pm \,:\, \dim(W \cap U)_{\{L\}} \geq r + 1 \}.$$

Thus V^r can be regarded as an orthogonal degeneracy locus studied in Chapters 6 and 7. Let us equip V^r with the scheme structure as in these chapters (i.e. induced from that of an appropriate Schubert scheme in an orthogonal Grassmann bundle). It is shown in [DC-P] that for a general curve C and any irreducible double cover, this scheme structure agrees on $V^r \smallsetminus V^{r+2}$ with the one defined by Welters stated in the theorem. Moreover, it makes V^r a reduced Cohen-Macaulay and normal scheme. We now want to apply a formula from the previous chapter to compute $[V^r]$. Fortunately, the Chern classes of the bundles W and U are very simple. Namely, using the Poincaré formula, one gets that

$$c_i(W^\vee) = \frac{(\theta')^i}{i!},$$

where θ' is the restriction of the class of the theta divisor on $\mathrm{Pic}^{2g-2}(\widetilde{C})$ to $P^{\pm}$. Moreover, $c_i(U) = 0$ for $i \geq 1$. Thus applying to our situation the formula (7.7) from the previous chapter, we must calculate

$$\widetilde{P}_{\rho(r)}(W^{\vee}) = 2^{-r}\widetilde{Q}_{\rho(r)}(W^{\vee}).$$

The element $\widetilde{Q}_{\rho(r)}(W^{\vee})$ is a certain rational multiple m of $(\theta')^{r(r+1)/2}$ obtained by specializing the Schur Pfaffian $\pi_{\rho(r)}(c)$ with $c_i = \frac{1}{i!}$. A general $\pi_{\lambda}(c)$, where $\lambda = (\lambda_1 > \ldots > \lambda_l > 0)$, becomes under this specialization

$$\frac{1}{\lambda_1! \ldots \lambda_l!} \prod_{p<q} \frac{\lambda_p - \lambda_q}{\lambda_p + \lambda_q}.$$

Indeed, once it is known that

$$\pi_{i,j}(c) = \frac{1}{i!j!}\frac{i-j}{i+j},$$

this equality is an immediate consequence of standard properties (D.3) and (D.6) of Pfaffians in Appendix D; see also [DC-P].

In our case, we get by an easy induction on r that

$$m = \prod_{i=1}^{r}\frac{1}{i!} \prod_{1 \leq j < i \leq r} \frac{i-j}{i+j} = \prod_{i=1}^{r} \frac{(i-1)!}{(2i-1)!}.$$

Invoking now (see [A-C-G-H, App.C] and [Mu2]) that the Prym variety is a principally polarized variety with the class ξ of the theta divisor equal to $\theta'/2$, we get the following formula for $[V^r]$. Suppose that V^r is either empty or of pure codimension $r(r+1)/2$. Then its class in $H^*(P^{\pm}, \mathbb{C})$ is equal to

$$2^{r(r-1)/2} \prod_{i=1}^{r} \frac{(i-1)!}{(2i-1)!}\xi^{r(r+1)/2}.$$

Using another induction on r, the coefficient of $\xi^{r(r+1)/2}$ can be written in a more elegant form:

$$2^{r(r-1)/2} \prod_{i=1}^{r} \frac{(i-1)!}{(2i-1)!} = \prod_{i=1}^{r}(2i-1)^{i-r-1}.$$

Therefore, we finally get the following answer:

$$[V^r] = \frac{1}{1^r \cdot 3^{r-1} \cdot 5^{r-2} \cdot \ldots \cdot (2r-1)}\xi^{r(r+1)/2}.$$

Invoking the ampleness of the theta divisor on $P^{\pm}$,[1] we get the following "existence" result. If $g \geq r(r+1)/2 + 1$, then V^r is not empty and of dimension at

---

[1]A divisor on X is ample if a certain multiple of it defines an embedding of X in some projective space. As is known from the theory of abelian varieties, 3 times any theta divisor defines an embedding into a projective space, hence such a divisor is ample. Any power of the cohomology class of an ample divisor on X, that is smaller than or equal to $\dim(X)$, is nonzero (see e.g. [F1, §12]). A cycle with a nonzero cohomology class must have a nonempty support (of suitably big dimension).

least $g - 1 - r(r + 1)/2$. This existence theorem had been originally obtained by Bertram using other methods.

For example, $[V^1] = \xi$, $[V^2] = \frac{1}{3}\xi^3$, $[V^3] = \frac{1}{45}\xi^6$, $[V^4] = \frac{1}{3^3 5^2 7}\xi^{10}$, $[V^5] = \frac{1}{3^4 5^3 7^2 9}\xi^{15}$.

The class of V^3 had been computed by Debarre using different tools. The formula for $[V^r]$ was first obtained in [DC-P] as in this chapter. It can be also deduced from the determinantal formula (6.5) from Chapter 6 – see [F4].

CHAPTER IX

APPLICATIONS AND OPEN PROBLEMS

Section 9.1 Components of degeneracy loci

In Chapter 1 we described which degeneracy loci are irreducible (for generic matrices), and gave some examples of rank conditions that do not give irreducible loci. B. Sturmfels has raised the interesting question of giving an algorithm for writing an arbitrary locus given by rank conditions on upper left submatrices as a union of its irreducible components.

Section 9.2 Existence and connectedness theorems

The simplest application of any of these formulas is to show that some locus Ω_w must be *nonempty*. One will know this as soon as one proves that the corresponding polynomial P_w is nonzero in $H^{2d}(X)$. Sometimes this can be done by direct calculation. One of the first important applications of this idea was the application of the Giambelli-Thom-Porteous formula by Kempf and Kleiman and Laksov to prove the existence of special divisors on a curve, when the expected dimension of this locus is nonnegative. Similarly, the formula discussed in Chapter 8 gives another proof of a theorem of Bertram, that the locus of special divisors on a Prym variety is nonempty when its expected dimension is nonnegative.

Formulas for degeneracy loci in the symplectic case (in both forms of Chapters 6 and 7) have been used by Ekedahl and van der Geer [vdG] to compute cycles on moduli spaces of abelian varieties.

The nonemptyness of a locus when the expected dimension is nonnegative has become part of a more general story. When the expected dimension is strictly positive, in the presence of suitable positivity of a variety or bundle, one expects more: that the locus must be *connected*. There are several notions of positivity for a vector bundle (the choice of which does not much affect the discussion here). The quickest to define is called ampleness. A line bundle L on X is ample if there is a positive integer m so that $L^{\otimes m}$ is the restriction to X of the line bundle $\mathcal{O}(1)$ for an embedding of X in some projective space. The bundle E on X is **ample** if, on the projective bundle $\mathbb{P}^*(E)$ of hyperplanes in E, the tautological quotient line bundle $\mathcal{O}(1)$ is ample.

In [F-L1] such a theorem was proved in the Thom-Porteous setup. There it was shown that for any map $h : F \to E$ of bundles on an irreducible variety, if $H = \mathrm{Hom}(F, E) = F^\vee \otimes E$ is an ample bundle, then the degeneracy locus $D_r(h)$ is connected whenever its expected dimension is positive, i.e., whenever $\dim(X) > (\mathrm{rank}(E) - r)(\mathrm{rank}(F) - r)$.

It is reasonable to hope for more results in this direction. For example, with the same assumptions, but with E and F each (partially) filtered, and w a permutation such that the degeneracy locus Ω_w is defined, then, if $l(w) < \dim(X)$, one may expect that Ω_w is connected. This would follow from a general conjecture made in [F-L2] that the intersection of an irreducible cone in an ample bundle by a section must be connected, when the expected dimension is positive. Perhaps the methods of [F-L1] can be used in this special case.

For symmetric and skew-symmetric bundle maps, Tu [T], Harris and Tu [H-T2], and Laytimi [Lay] have proved some of the expected connectedness formulas, although even here the story is not yet complete. Are there analogous assertions for other degeneracy loci, under appropriate positivity assumptions on the bundles V and L involved in the bilinear form $V \otimes V \to L$?

Section 9.3 Positivity

One of the lessons that should be learned from this story is that these Schur and Schubert polynomials are fundamental polynomials. The Schur polynomials $s_\lambda(x)$ form a basis for the symmetric polynomials, that most would agree should be known to every mathematician with any interest in algebra or geometry. The Schubert polynomials are a similarly fundamental collection of polynomials, that give a basis for all polynomials.

For a vector bundle E of rank r on a projective variety X, and any symmetric polynomial $P(x_1, \ldots, x_r)$ that is homogeneous of degree d, one can evaluate P on the Chern classes of E, by writing P as a polynomial in the elementary symmetric polynomials $e_i(x)$, and substituting $c_i(E)$ for $e_i(x)$. Denote the resulting class $P(E)$ in $H^{2d}(X)$. If V is any d-dimensional (irreducible, closed) subvariety V of X, one can cap this class $P(E)$ with the fundamental homology class $[V]$ of V in $H_{2d}(X)$, obtaining a class in $H_0(X)$. The degree of this class is often denoted $\int_V P(E)$, since it can be realized as an integral if the cohomology class is represented by a differential form.

Griffiths asked in 1969 which polynomials P have the property that this number $\int_V P(E)$ is always positive, whenever the bundle E is ample. Call such polynomials **positive**. Some cases were known, written in terms of Chern classes, i.e., of elementary symmetric polynomials in the x variables: c_1 is positive, c_2 is positive (this is harder), $c_1{}^2 - c_2$ is positive. (One red herring: by a miscalculation, it was thought that $c_1{}^2 - 2c_2$ is positive, but it is not. Griffiths conjectured a general answer, which was in fact correct, but it was — as seen by this example — not easy to calculate!) All c_n were known to be positive (Bloch and Gieseker), although no proof is known even today that doesn't use the Hard Lefschetz Theorem. Some others, such as $c_1 c_2 - c_3$ and $c_1{}^3 - 2c_1 c_2 + c_3$, were known to be positive. Of course, any positive linear combination of positive polynomials is positive.

Note that if E is generated by its sections — a very special extra assumption — then E is the pullback of the universal quotient on a Grassmannian. We know that the corresponding classes of Schubert varieties are nonnegative, and, by Giambelli, we know how to express these as polynomials. S. Usui and H. Tango pointed out that these classes were positive in this special case. This quotient bundle is not ample, in fact, but it can be regarded as "almost ample." At least, it leads to the right guess, that was found and proved in 1982. Suppose P is a nonzero polynomial, homogeneous of degree d. Write $P = \sum a_\lambda s_\lambda$, for (unique) rational numbers a_λ, the sum over partitions λ of d in at most r parts, where r is the rank of the bundles under consideration.

THEOREM. [F-L2] P *is positive if and only if each a_λ is nonnegative.*

There is a natural generalization. Suppose E is a bundle together with a filtration $E_1 \subset \cdots \subset E_s \subset E$ of subbundles, of certain ranks. In this case one can define $P(E)$ when P is any polynomial in the Chern roots, numbered from top to bottom, provided it is symmetric in x_i and x_{i+1} whenever $i \notin T$, where T is the set of ranks of the quotient bundles E/E_i, $1 \le i \le s$. When is such a nonzero polynomial P positive? Anyone following these lectures should not hesitate very long to guess the answer! Write $P = \sum a_w \mathfrak{S}_w$, the sum over permutations w such that $w(i) < w(i+1)$ for all i not in T.

THEOREM. [F5] P *is positive whenever E is ample if and only if each a_w is nonnegative.*

Such polynomials include products of Schur polynomials of the quotient bundles E/E_i, which were known to be positive by the proof of the preceding theorem. However, there are new polynomials that are not in the cone generated by these. Joe Harris points out that this can be used to give an obstruction for the "extension problem" of finding an ample bundle E with given graded pieces E_i/E_{i-1}.

The proof is a combination of two things. One is the theorem stated in Chapter 2. This is applied to a "universal" case of the theorem, on a space $H = \text{Hom}(V, E)$, with V a trivial bundle, with a trivial filtration. On H there is a tautological map, whose degeneracy loci Ω_w are represented by $\pi^*(\mathfrak{S}_w(E))$, where $\pi : H \to X$ is the projection. If $s : X \to H$ is the zero section, then $s^*[\Omega_w] = \mathfrak{S}_w(E)$. Now H is an ample bundle, and it is a general fact — the main tool in [F-L2] — that the pullback of any subcone of an ample vector bundle by a section is positive.

To prove the converse, if one takes a polynomial in which a coefficient a_w is negative, one must construct a filtered ample bundle E on some variety X with $\int_V P(E) < 0$ for some subvariety V. The universal flag on an appropriate partial flag variety will have this property, with V the Schubert subvariety dual to Ω_w. This bundle is not quite ample, but, as in [F-L2], one can find a finite map $f : V' \to V$, with an ample line bundle L on V', such that $f^*(E) \otimes L$ is the desired ample bundle.

Note that the questions had nothing to do with Schur or Schubert polynomials. The answers give another indication of their fundamental role in geometry.

Demailly, Peternell, and Schneider have extended the result of [F-L2] to bundles that are only known to be nef, on a complex manifold [D-M-P]. When E is ample *and generated by its sections*, Beltrametti, Schneider, and Sommese [B-S-S] conjecture that, for P symmetric and positive as above, and E ample,

$$\int_V P(E) \ge P(1, 1, \dots, 1).$$

For example, if $X = \mathbb{P}^n$, and $E = \mathcal{O}(1)^{\oplus r}$, equality holds. It is natural to conjecture that the same is true in the filtered case.

There are some interesting applications of this geometry to algebra, something like what we saw about the positivity of the coefficients of the product of Schubert polynomials. From any vector bundle E one can construct new bundles, such as the symmetric or exterior powers $S^n E$ and $\wedge^n E$. More generally, for each partition μ there is a bundle $S^\mu E$ (see Appendix I). Take any positive polynomial P in the Chern roots of $S^\mu E$. One knows, at least in principle, how to write the Chern classes of $S^\mu E$ in terms of the Chern classes of E, so one can write $P = \sum a_\lambda s_\lambda$ for some unique coefficients a_λ, determined purely algebraically from P.

COROLLARY. ([P4, Cor. 7.2]). *These coefficients a_λ are all nonnegative.*

The argument is based on the fact that if E is ample, then $S^\mu E$ is also ample. The result then follows from the theorem. There are extensions of this theorem that use the filtered version of the theorem. Note that, so far, there is no proof of this corollary without using the Hard Lefschetz Theorem!

In case $c(E^\vee \otimes M) = c(F^\vee \otimes M) = c(G)$, where G is some ample vector bundle on X, then the polynomials occurring in the formulas (6.5)–(6.7) for the loci D_k are all multiples of Schur polynomials in the Chern classes of G, so they are numerically positive. In this case one gets an easy proof that the loci D_k must be nonempty.

Section 9.4 K-theory

For the (A_n) case there are analogous formulas in K-theory, i.e., in the Grothendieck ring of vector bundles. Assume for simplicity that the base variety X is nonsingular, so that every coherent sheaf $\mathcal{L}$ determines a class $[\mathcal{L}]$ in $K(X)$. The classes of structure sheaves $[\mathcal{O}_{\Omega_w}]$ still give an additive basis for $K(\mathcal{F})$ over $K(X)$, where $\mathcal{F}$ is a flag bundle as before, and Ω_w is defined as in §2.3. With notation as in that section, let a_i and b_i be the classes in $K(\mathcal{F})$ defined by the line bundles U_{n-i+1}/U_{n-i} and F_i/F_{i-1} respectively. Again $K(\mathcal{F})$ is generated over $K(X)$ by $a_1, \ldots, a_n$, with relations generated by all $e_i(a_1, \ldots, a_n) - e_i(b_1, \ldots, b_n)$. One can again ask for formulas to express the classes of these structure sheaves in terms of the algebraic generators. Lascoux [Las1] has given polynomials that he calls "Grothendieck polynomials," such that $[\mathcal{O}_{\Omega_w}] = G_w(a; b)$. In fact, the construction, and the proof of this formula, are very much like that given in Chapter 2; cf. [F-Las].

Stating these formulas in K-theory raises the question: Can one find explicit resolutions of these structure sheaves by locally free sheaves, from which one can read off these formulas in K-theory? These complexes would be fundamental universal complexes, one for each permutation w, generalizing the Koszul complexes and Eagon-Northcott complexes familiar in commutative algebra. Little seems to be known about this beyond the case on the Grassmannian studied by Lascoux in his thesis two decades ago.

One can ask questions about these classes similar to those which were considered for the cohomology classes in earlier chapters. For example, one can look for formulas for their products. We mention one case of this, proved in a more general setting in [F-Las], since it implies some unexpected positivity. There are some (relatively) ample line bundles on these flag varieties; in the absolute case when X is a point, they are tensor products of pullbacks of the ample line bundles on Grassmannians that come from the Plücker embeddings. One can ask for the class of the restriction of such a line bundle L to a Schubert variety Ω_w, as expanded in the basis $[\mathcal{O}_{\Omega_v}]$. Only those Ω_v that are contained in Ω_w occur, but what is surprising is that the coefficient of each such class is *nonnegative*, i.e., a nonnegative linear combination of classes of line bundles on X. On a Grassmannian, for example, with its canonical ample line bundle L, our formula gives the identity

$$[L\,|_\Omega] = \sum [\mathcal{O}_{\Omega'}],$$

the sum over all Schubert subvarieties Ω' of Ω; each occurs with the coefficient 1.

This raises many questions, starting with the question of understanding a reason for this, other than a complicated calculation. The fact itself explains, and is closely related to, some questions about "postulation," or "standard monomial theory," which means finding dimensions of, or a basis for, sections of such line bundles. For example, it follows from the displayed formula that the dimension of the sections of $L^{\otimes r}$ on a given Schubert variety Ω is the number of chains of length r of Schubert subvarieties: $\Omega = \Omega(0) \supseteq \Omega(1) \supseteq \cdots \supseteq \Omega(r)$, a formula that goes back to Hodge in case Ω is the Grassmannian. Note that the K-theory formula on the Grassmannian refines a simpler formula in geometry: the hyperplane intersection $c_1(L) \cap [\Omega]$ is the sum of all $[\Omega']$, for those $\Omega' \subset \Omega$ of codimension one.

This standard monomial theory is now quite well developed for the classical groups, by Seshadri, Lakshmibai, Musili, and others, and related combinatorics is also developing. However, we do not know of analogues of the results of [F-Las]. For example, is there a corresponding positivity result for other Grassmannians or flag varieties? Is there a general group-theoretic approach?[1]

Section 9.5 Schubert polynomials for classical groups

A major project, which was one of the motivations for this Thurnau Summer School, is to understand the relations among the various approaches to Schubert polynomials for the other classical groups. It has become clear, and pointed out explicitly by Fomin and Kirillov [F-K], that one cannot hope for Schubert polynomials for other cases that simultaneously have all the wonderful properties of those of Lascoux and Schützenberger for (A_n). They show, even for the case of (B_2), one cannot define polynomials of the right degrees, that represent the cohomology classes in the flag manifold, that are expressed as nonnegative combinations of monomials in the basic classes x_i, and have the property that products of these polynomials are equal to the same linear combination of other classes as one sees for the products of cohomology classes. The polynomials of [B-H] have positive expansions in terms of some other variables. In [F-K] others are constructed, including some simple polynomials that multiply correctly but don't represent the cohomology classes. The formulas we have discussed in Chapters 6 and 7 produce some other candidates; using these, one will have to at least give up the property that their multiplication is the same as that of cohomology classes.

Another desirable property of Schubert polynomials is that they should be part of a general story for *double* Schubert polynomials. Indeed, we have seen that double

---

[1]Such a positivity theorem has recently been proved by O. Mathieu, using character formulas.

Schubert polynomials are the natural polynomials to repressent degeneracy loci. We would like any candidates for double Schubert polynomials to give formulas for degeneracy loci, which means that they must equal the classes described here in the corresponding ring modulo the relations.

Where several approaches or formulas have appeared, we would like to understand the relations among them better. For example, we would like an algebraic proof that the determinantal formulas from Chapter 6 agree with those from Chapter 7 as sums of products of $\widetilde{Q}$ and $\widetilde{P}$ polynomials, i.e., that they are the same in the corresponding ring $\mathbb{Z}[x_1, \ldots, x_n, y_1, \ldots, y_n]/I$, where I is generated by differences of appropriate symmetric polynomials in the x's and y's, as in Chapter 6.[2]

In general, one would like more explicit formulas, such as the various Schur-type formulas for the classical Schubert polynomials for vexillary permutations. The determinantal formulas emphasized in Chapter 6 should be a small part of this story, as well as the formulas on Grassmann bundles discussed in Chapter 7.

Section 9.6 Exceptional groups

The original story of [B-G-G] and [D2] was for arbitrary semisimple Lie groups. Are there analogues of the story developed here for the five exceptional groups? What is the global setup, corresponding to the bundle V with its symplectic or quadratic form? Presumably, it is a bundle V with some additional structure, such as a trilinear form in the case of (G_2). This should be an interesting project for someone with an interest in exceptional groups and geometry.

Section 9.7 Intersection theory refinements

There has been a lot of work squeezing more information out of, and proving refined versions of, Bézout's theorem. These aim at saying something about the intersection of hypersurfaces when they do not meet properly, i.e., the intersection is bigger than expected. The first and simplest such statement, that grew out of collaboration with MacPherson and Lazarsfeld [F1], says that the sum of the degrees of the irreducible components of the intersection is always less than or equal to the product of the degrees of the hypersurfaces. This has been reproved and refined in various ways by Vogel and his colleagues. It is natural to expect that there would be analogues of these refinements for more general degeneracy loci.

Section 9.8 Reality

Formulas for degeneracy loci predict the number of solutions of certain equations (or of certain geometric figures) over an algebraically closed field, usually $\mathbb{C}$. One can ask to what extent these can be realized over other fields, such as $\mathbb{R}$ (or even $\mathbb{Q}$ or finite fields). For example, in Grassmannians or flag varieties, products of Schubert polynomials tell how many flags are in given attitude with a collection of flags in general position. Can one find such a collection of real flags so that all the predicted complex solutions are real? Surprisingly, we have been unable to locate any classical reference to this problem, although the necessity of extending from

---

[2]This problem has recently been solved in [L-P].

$\mathbb{R}$ to $\mathbb{C}$ was certainly well understood. (Classical mathematicians were certainly interested in and aware of similar problems, such as the number of real flexes on a real plane cubic, where the real answer is smaller than the complex answer.) F. Sottile, in his 1994 University of Chicago thesis (see [S2]), showed that the answer is positive for the Grassmannian of 2-planes in n-space, for any n, and for some other cases. Are there corresponding results in other flag varieties and partial flag varieties, or for other degeneracy loci for matrices of forms as in Chapter 1?

Section 9.9 Arithmetic Schubert calculus

Arakelov originated an intersection theory for arithmetic surfaces, using some analysis at infinite places to go with algebraic geometry at finite places. Gillet and Soulé extended this theory to higher dimensions. A Schubert calculus for arithmetic Grassmannians was begun by Maillot [Mai] and continued by Tamvakis [Ta1], [Ta2], who also gave extensions to partial flag varieties. Bost, Gillet, and Soulé [B-G-S] have found analogues of Bézout's theorem in the arithmetic setting. One may expect that many formulae for degeneracy loci will have arithmetic realizations.

Section 9.10 Intersection products in flag manifolds

In Chapter 2 we mentioned the Littlewood-Richardson rule for multiplying Schubert varieties in a Grassmannian, or multiplying Schur polynomials. As of today, although many would claim that the cohomology ring of a flag manifold is well understood, there is no analogue of the Littlewood-Richardson rule for explicitly multiplying Schubert classes in a flag manifold. It is a fact that the Schubert polynomials $\mathfrak{S}_w$ are a basis for all polynomials, so there are formulas

$$\mathfrak{S}_u \cdot \mathfrak{S}_v = \sum c_{uv}^w \, \mathfrak{S}_w,$$

for any permutations u and v, where the sum is over all w (perhaps in a larger symmetric group) with $l(w) = l(u) + l(v)$, and the coefficients are integers. By looking at the corresponding intersections in a (suitably large) flag variety, one can see that each of these coefficients must be nonnegative, but one does not know a formula for them. We have seen Monk's formula, which is the case where the length of u is 1, generalizing Pieri's formula on a Grassmannian. Lascoux and Schützenberger stated a generalization of Monk's formula, which, in another form, can be found, with a proof, in Sottile [S1].

There are some analogues of Pieri's formula for the other classical Grassmannians (cf. Chapters 3 and 7 and Appendix E), but, needless to say, the story is even less complete than in the (A_n) case.

INTERSECTION THEORY; POLYNOMIALS REPRESENTING DEGENERACY LOCI

A.1 Cycle classes

In this text, for simplicity, we generally work with complex varieties. We require some basic facts about homology and cohomology classes determined by subvarieties of complex varieties. One reference for how to carry this out is Appendix B of [F6]. Those familiar with the more refined Chow groups and their intersection theory [F1] should have no difficulty using these; with this, the results are valid for varieties or schemes over arbitrary fields, at least if characteristic 2 is avoided when dealing with quadratic forms.

In this appendix we state what is needed from these areas. The assertion that a polynomial represents a degeneracy locus may also be given several interpretations, depending on the reader's background, and we discuss various cases here.

The cohomology groups form a graded ring, with the cup product[1] $\cup$, and the homology groups form a module over the cohomology ring, by means of the cap product $\cap$. When X is a nonsingular complex projective variety of dimension n, it is an oriented real $2n$-manifold, and the group $H_{2n}(X)$ has a canonical generator $[X]$; the canonical map $H^i(X) \to H_{2n-i}(X)$, taking α to $\alpha \cap [X]$, often called the Poincaré duality map, is an isomorphism.

Since homology is covariant, and cohomology contravariant, we have pushforward maps $f_*: H_i(X) \to H_i(Y)$ and pullback maps $f^*: H^i(Y) \to H^i(X)$ for a continuous map $f: X \to Y$. These are related by the projection formula $f_*(f^*(\alpha) \cap \beta) = \alpha \cap f_*(\beta)$ for $\alpha \in H^i(Y)$ and $\beta \in H_j(X)$. If X and Y are nonsingular and projective, and one identifies homology with cohomology, one gets pushforward maps $f_*: H^i(X) \to H^{i+2d}(Y)$, where $d = \dim(Y) - \dim(X)$.

A (closed) subvariety V of a projective variety X of dimension k determines a class denoted $[V]$ in $H_{2k}(X)$, where k is the dimension of V. This can be extended to arbitrary subschemes V of pure dimension k, by setting $[V] = \sum m_i[V_i]$, if V_i are the irreducible components of V and m_i is the multiplicity of V_i (i.e., the length of the local ring of V at the generic point of V_i). If X is nonsingular, by Poincaré duality, we have the class $[V]$ in $H^{2c}(X) = H_{2k}(X)$, where c is the codimension of V in X.

If $f: X \to Y$ is a morphism of projective varieties, and V is a subvariety of X, with $W = f(V)$, then $f_*[V] = 0$ if $\dim(W) < \dim(V)$, and $f_*[V] = d \cdot [W]$ if the map from V to W is generically d to 1; in particular, $f_*[V] = [W]$ if f maps V

---

[1]Often the cup product $c \cup d$ is denoted by $c \cdot d$ or cd.

birationally onto W. If f is a smooth morphism from X to Y (such as the projection of a projective bundle), and V is a subvariety of Y, then $f^*[V] = [f^{-1}(V)]$.

Subvarieties V and W of a smooth projective variety X are said to meet transversally if their intersection is a union of subvarieties $Z_1, \ldots, Z_r$, with $\mathrm{codim}(Z_i, X) = \mathrm{codim}(V, X) + \mathrm{codim}(W, X)$ for each i, and the tangent space to Z_i at a general point is the transversal intersection of the tangent spaces to V and to W at that point. In this case $[V] \cup [W] = [Z_1] + \cdots + [Z_r]$.

If c is any cohomology class on an irreducible projective variety X, the notation $\int_X c$ is often used to denote the image of $c \cap [X]$ by the degree mapping (push-forward) $H_*(X) \to H_*(\mathrm{pt.}) = \mathbb{Z}$. The intersection number $\int_X [V] \cup [W]$ of two subvarieties V and W of complementary dimension is often denoted $\langle V, W \rangle$.

If a projective X has a filtration $X = X_s \supset X_{s-1} \supset \cdots \supset X_0 = \emptyset$ by closed algebraic subsets, and each $X_i \smallsetminus X_{i-1}$ is a disjoint union of varieties $U_{i,j}$ each isomorphic to an affine space $\mathbb{C}^{n(i,j)}$, then the classes $[\overline{U}_{i,j}]$ of their closures give a basis for $H_*(X) = \oplus H_i(X)$ over $\mathbb{Z}$. (In particular, all odd groups $H_{2i+1}(X)$ must vanish.) These conditions apply to Schubert varieties in flag manifolds, which are the closures of Schubert cells.

It is also a general fact that if a connected algebraic group G acts on a variety X, the corresponding action on the cohomology is trivial, so $[g \cdot V] = [V]$ for a subvariety V and an element g in G. If G acts transitively on X, one can use this to make $g \cdot V$ meet a given subvariety W transversally.

A vector bundle E of rank r on a variety X has Chern classes $c_i(E)$ in $H^{2i}(X)$, with $c_0(E) = 1$ and $c_i(E) = 0$ if $i > r$. If $f: Y \to X$ is a morphism, then $c_i(f^*(E)) = f^*(c_i(E))$. These classes satisfy the Whitney formula: if E' is a subbundle of E, with quotient bundle E'', then $c_i(E) = \sum_{j+k=i} c_j(E') \cup c_k(E'')$. If X is nonsingular, and s is a section of E that is transversal to the zero section, then $c_r(E) = [Z(s)]$, where $Z(s)$ is the zero locus of s.

These results can be extended to noncompact varieties, but then another homology theory (with locally finite chains, called Borel-Moore homology) must be used. The Borel-Moore homology groups of a variety X can be defined by setting $H_k(X)$ to be $H^{N-k}(M, M \smallsetminus X)$, where M is any oriented N-dimensional manifold containing X as a closed subset. Indeed, this generalization provides one of the easiest ways to prove the basic properties (see [F6] for details). An alternative, which is valid over any field, and gives finer results, is to use Chow groups $A_k(X)$ instead of $H_{2k}(X)$; with these there are Chow cohomology groups $A^c(X)$ to use instead of $H^{2c}(X)$. (See [F1] for this.)

In most of this text, it does not matter very much which of these theories is used to carry cycle classes. There are some situations, however, where their differences influence the discussion. These differences show up most vividly when Y is a closed subvariety (or subscheme) of an algebraic variety X, and U is the complement of Y in X. All three theories have pushforward maps for the inclusion of Y in X. Only the Chow theories and Borel-Moore homology theories have restriction (pullback) maps for the open inclusion of U in X, however. For the Chow groups these maps determine short exact sequences

$$A_k(Y) \to A_k(X) \to A_k(U) \to 0.$$

For Borel-Moore homology, one has a long exact sequence

$$\cdots \to H_{k+1}(U) \to H_k(Y) \to H_k(X) \to H_k(U) \to H_{k-1}(Y) \to \cdots \quad .$$

A.2 Polynomials representing degeneracy loci

We want to say now what it means for a polynomial (in some Chern classes of some vector bundles) to represent a degeneracy locus. The essential case can be seen in the situation just discussed: the zero locus $Z(s)$ of a section s of a vector bundle E of rank r is represented by the top Chern class $c_r(E)$. This statement can have several meanings, depending on the taste and sophistication of the reader.

(1) If X is nonsingular and projective, and s is sufficiently generic, then $[Z(s)] = c_r(E)$.

In this case, being transversal to the zero section is sufficiently generic. For a reader new to intersection theory, this meaning suffices, and the entire text can be read with only this interpretation. However, some intersection theory allows several improvements.

(2) If X is a Cohen-Macaulay variety (or scheme of pure dimension n), and $\mathrm{codim}(Z(s), X) = r$, then $[Z(s)] = c_r(E) \cap [X]$ in $H_{2n-2r}(X)$,

where $[Z(s)]$ is the class of the scheme $Z(s)$ with its natural scheme structure (defined locally by the r functions given by the section of a trivial bundle).

Without any genericity assumptions, it is a general fact that the codimension of each component of $Z(s)$ in X is at most r. There is always a class, denoted $\mathbf{Z}(s)$, in $H_{2n-2r}(Z(s))$, where X is a scheme of pure dimension n; the image of $\mathbf{Z}(s)$ in $H_{2n-2r}(X)$ is $c_r(E) \cap [X]$. This class $\mathbf{Z}(s)$ has several properties. With the assumptions of (2), $\mathbf{Z}(s) = [Z(s)]$. More generally, $\mathbf{Z}(s) = [Z(s)]$ whenever $\mathrm{depth}(Z(s), X) = r$. In general we have the following statement:

(3) If X has pure dimension n, and each component of $Z(s)$ has its expected dimension $n - r$, then $\mathbf{Z}(s)$ is a positive cycle whose support is $Z(s)$.

The class $\mathbf{Z}(s)$ can be constructed as follows. One has a Thom class Θ in the relative cohomology group $H^{2r}(E, E \smallsetminus 0)$, where $0 \subset E$ is the image of the zero section from X to E. Then $s^*(\Theta)$ is in $H^{2r}(X, X \smallsetminus Z(s))$, and the cap product of $s^*(\Theta)$ with the fundamental class $[X]$ in $H_{2n}(X)$ is the class $\mathbf{Z}(s)$ in $H_{2n-2r}(Z(s))$. In [F1] a construction of a refined class, in the Chow group $A_{n-r}(X)$, is denoted $s^!(0)$.

(4) The formation of $\mathbf{Z}(s)$ is compatible with basic constructions in intersection theory: proper pushforward, flat pullback, and pullback by a local complete intersection morphism.

For example if $f \colon Y \to X$ is a projection of a bundle, then $\mathbf{Z}(f^*s) = f^*(\mathbf{Z}(s))$. In fact, one can refine this one step further, to a class $\mathbf{Z}'(s)$ in the local cohomology group $H^{2r}_{Z(s)}(X) := H^{2r}(X, X \smallsetminus Z(s))$, which is also compatible with the basic constructions of intersection theory. This is the class $s^*(\Theta)$ considered above. In the corresponding Chow theory of algebraic cycles, this refinement lies in what is called the bivariant Chow group $A^r(Z(s) \to X)$. We refer to [F1] for more about this.

There are similar meanings for a polynomial to **represent a degeneracy locus**. Consider the situation of Theorem 2 in Chapter 2, where one has a morphism h of flagged vector bundles, with rank conditions r. In the simplest case, if X is a Cohen-Macaulay (e.g., smooth) complex scheme of pure dimension n, and the codimension of $\Omega_r(h)$ is the expected number $d(r) = |\lambda| = \sum m_i p_i$, then the fundamental class $[\Omega_r(h)]$ of this scheme $\Omega_r(h)$ is equal to $P_r \cap [X]$ in $H_{2n-2d(r)}(X)$, where P_r is the polynomial in the theorem. Without these genericity assumptions, if X has pure dimension n, the codimension of each component of $\Omega_r(h)$ is at most $d(r)$. There is always a class, denoted $\mathbf{\Omega}_r(h)$, in the homology group $H_{2n-2d(r)}(\Omega_r(h))$, whose image in $H_{2n-2d(r)}(X)$ is $P_r \cap [X]$. The formation of this class $\mathbf{\Omega}_r(h)$ commutes with flat pullback, local complete intersection pullback, or push-forward by proper morphisms. If $\Omega_r(h)$ has pure codimension $d(r)$, then $\mathbf{\Omega}_r(h)$ is a positive cycle whose support is exactly $\Omega_r(h)$. These classes can also be refined to classes in local cohomology groups or bivariant groups; in fact, $\mathbf{\Omega}_r(h) = [\Omega_r(h)]$ if $\operatorname{depth}(\Omega_r(h), X) = d(r)$ (cf. [F1, Ex. 14.3.1]).

In general these classes $\mathbf{\Omega}_r(h)$ are constructed by taking the refined pullback of the class $[\Omega_r]$ of a locus Ω_r constructed in a universal situation, such as a flag bundle $\mathcal{F}$ over X, by a section s. This locus $\Omega_r \subset \mathcal{F}$ has the expected dimension, and, with its natural scheme structure, is reduced, if X is reduced. We use the Poincaré-Lefschetz-Alexander isomorphism: for a (decent) closed subspace Z of a compact oriented real N-manifold M,

$$H_k(Z) \cong H^{N-k}(M, M \smallsetminus Z).$$

Let b be the dimension of the fibers of $\mathcal{F}$ over X. The dimension of Ω_r is $n+b-d(r)$, so we have

$$[\Omega_r] \in H_{2(n+b-d(r))}(\Omega_r) \cong H^{2d(r)}(\mathcal{F}, \mathcal{F} \smallsetminus \Omega_r).$$

Now the pullback by s is a map

$$s^* : H^{2d(r)}(\mathcal{F}, \mathcal{F} \smallsetminus \Omega_r) \to H^{2d(r)}(X, X \smallsetminus \Omega_r(h)) \cong H_{2n-2d(r)}(\Omega_r(h)),$$

the last equality again by duality. The image of $[\Omega_r]$ in $H_{2n-2d(r)}(\Omega_r(h))$ is defined to be the class $\mathbf{\Omega}_r(h)$. This can all be extended to varieties that are not complete, by using Borel-Moore homology instead of singular homology.

Over an arbitrary field, one may use the Chow groups, defining $\mathbf{\Omega}_r(h) = s^![\Omega_r]$, where $s^!$ is the refined pullback

$$s^! : A_i(\Omega_r) \to A_{i-b}(\Omega_r(h))$$

determined by the regular embedding $s : X \to \mathcal{F}$ of codimension b. In this case, the class $s^![\Omega_r]$ is constructed by deformation to the normal bundle, from which the four properties are proved. Property (3) is a general fact about local complete intersections in Cohen-Macaulay schemes, first pointed out by Kempf and Laksov (cf. [F1, Prop. 7.1]). It follows from the following fact, whose proof we include since it has not always been clear in the literature.

LEMMA. *Let $f: X \to Y$ be a morphism from a pure-dimensional scheme X to a nonsingular variety Y and let V be a closed subscheme of Y that is Cohen-Macaulay and of pure codimension d in Y. Let $W = f^{-1}(V)$. Then $\operatorname{codim}(W, X) \leq d$. If X is Cohen-Macaulay and $\operatorname{codim}(W, X) = d$, then W is Cohen-Macaulay and $f^*[V] = [W]$.*

PROOF. Factor f into $p \circ i$, where $i: X \to X \times Y$ is the graph of f, and $p: X \times Y \to Y$ is the projection. By the definition of the pullback, $f^*[V] = i^*p^*[V] = i^*[X \times V]$. Now i is a regular embedding, since Y is nonsingular; and $i^{-1}(X \times V) = W$ by construction. The product of two Cohen-Macaulay schemes is Cohen-Macaulay, so $X \times V$ is Cohen-Macaulay if X is Cohen-Macaulay. The conclusions follow from a basic fact of commutative algebra: in a local Noetherian ring, a sequence of r elements in the maximal ideal cuts the dimension down by at most r; when the ring is Cohen-Macaulay, the dimension is cut down by precisely r exactly when the residue ring is Cohen-Macaulay and the sequence is a regular sequence. (See [F1, Lemma A.7.1 and Proposition 7.1].) □

The conclusion of this lemma can fail if X is not Cohen-Macaulay, even if Y and V are nonsingular, cf. [F1, Example 7.1.5].

We apply the lemma to the situation where X is a Cohen-Macaulay variety, $\mathcal{F}$ is a bundle over X, and $\Omega \subset \mathcal{F}$ is some "universal" degeneracy locus. One has a section $s: X \to \mathcal{F}$, and one wants to know that the codimension of $s^{-1}(\Omega)$ in X is at most the codimension of Ω in $\mathcal{F}$, that $s^{-1}(\Omega)$ is Cohen-Macaulay when the codimensions are equal, and that $s^*[\Omega] = [s^{-1}(\Omega)]$. This problem is local on X, and locally one finds a map $\mathcal{F} \to Y$ to a nonsingular variety Y (often a space of matrices or a flag variety), and one has a subvariety $\tilde{\Omega}$ of Y that one knows to be Cohen-Macaulay by some universal local study, and so that Ω is the pullback of $\tilde{\Omega}$ by the map from $\mathcal{F}$ to Y. One then applies the lemma to the composite map $X \to \mathcal{F} \to Y$ and the subvariety $V = \tilde{\Omega}$ of Y.

For example, in the Thom-Porteous situation, one takes $\mathcal{F} = \operatorname{Hom}(F, E)$, with Ω the locus of maps of rank at most r, and s the section corresponding to a vector bundle map φ from F to E on X. One deduces that the degeneracy locus $D_r(\varphi) = s^{-1}(\Omega)$ has codimension at most the expected codimension, and that, if X is Cohen-Macaulay, and the dimension is as expected, then $D_r(\varphi)$ is Cohen-Macaulay, and its class is the cap product of the Giambelli-Thom-Porteous polynomial with the fundamental class of X.

It is also true that the topological construction, giving a class in homology groups, is compatible with that using intersection theory. The class $\Omega_r(h)$ can be constructed in the bivariant Chow group $A^{d(r)}(\Omega_r \to X)$, which strengthens the Property (4).

An arbitrary polynomial P in Chern classes of some vector bundles on a variety X is said to be **supported** on a closed subscheme Y of X if $P \cap [X]$ is in the image of the map from $A_*(Y)$ to $A_*(X)$ For most purposes, one could use the weaker assertion that $P \cap [X]$ is in the image of the map $H_*(Y) \to H_*(X)$ in homology. A polynomial that represents a degeneracy locus, for example, is always supported on that locus.

VEXILLARY PERMUTATIONS AND
MULTI-SCHUR POLYNOMIALS

Lascoux and Schützenberger proved that if w is a **vexillary** permutation, which means that there is no

$$a < b < c < d \quad \text{with} \quad w(b) < w(a) < w(d) < w(c),$$

then the corresponding Schubert polynomial can be written as a certain Schur determinant. In [F2] we gave a characterization of vexillary permutations in a form suitable for degeneracy loci. In this appendix we improve this result, by allowing rank conditions that may be redundant.[1] Since a vexillary permutation can have several such descriptions, this gives several different determinantal formulas for their Schubert polynomials.

To describe this, we need some notation. For any nonnegative integers u and v, let

$$h_l(u, v) = \text{the coefficient of } T^l \text{ in } \prod_{i=1}^{u}(1 - y_i T) \Big/ \prod_{j=1}^{v}(1 - x_j T).$$

For any sequences $p_1 \geq p_2 \geq \cdots \geq p_k$ and $m_1, \ldots, m_k$ of nonnegative integers, let λ be the partition $(p_1{}^{m_1}, \ldots, p_k{}^{m_k})$, and set $m = m_1 + \cdots + m_k$. For any nonnegative integers $u_1, \ldots, u_k, v_1, \ldots, v_k$, define the **multi-Schur polynomial**

$$s_\lambda((u_1, v_1)^{m_1}, \ldots, (u_k, v_k)^{m_k})$$

to be the determinant of the $m \times m$ matrix $\big(h_{\lambda_i - i + j}(\bullet, \bullet)\big)$, where, in the first m_1 rows the dots are replaced by (u_1, v_1), and in the next m_2 rows the dots are replaced by (u_2, v_2), and so on for all the rows of the matrix. (Note that we allow some m_i to be zero, in which case there are no corresponding rows.)

PROPOSITION. *Let* $a_1, \ldots, a_k, b_1, \ldots, b_k$, *and* $r_1, \ldots, r_k$, *be nonnegative integers satisfying the inequalities*

$$
\begin{align}
&(i) & &a_1 \geq a_2 \geq \cdots \geq a_k; \\
&(ii) & &b_1 \leq b_2 \leq \cdots \leq b_k; \\
&(iii) & &a_1 - r_1 \geq a_2 - r_2 \geq \cdots \geq a_k - r_k \geq 0; \\
&(iv) & &0 \leq b_1 - r_1 \leq b_2 - r_2 \leq \cdots \leq b_k - r_k.
\end{align}
$$

---

[1] The possibility of such a generalization was discussed in Remark 9.16 of [F2]. This appendix, in fact, corrects the last two sentences of that remark.

For any $n \geq a_1 + b_k$ there is a unique w in S_n whose essential set is contained in the set $\{(a_1, b_1), (a_2, b_2), \ldots, (a_k, b_k)\}$, and with $r_w(a_i, b_i) = r_i$ for $1 \leq i \leq k$. The Schubert polynomial of w is given by

$$\mathfrak{S}_w(x, y) = s_\lambda((b_k, a_k)^{m_1}, \ldots, (b_1, a_1)^{m_k}),$$

where $\lambda = (p_1{}^{m_1}, \ldots, p_k{}^{m_k})$, with

$$p_i = b_{k+1-i} - r_{k+1-i} \text{ for } 1 \leq i \leq k, \quad \text{and}$$
$$m_1 = a_k - r_k, \quad m_i = (a_{k+1-i} - r_{k+1-i}) - (a_{k+2-i} - r_{k+2-i}) \text{ for } 2 \leq i \leq k.$$

In fact, the pairs (a_i, b_i) that are not in the essential set of w are exactly those such that either $b_i - r_i = b_{i-1} - r_{i-1}$ (or $b_1 - r_1 = 0$ if $i = 1$), or $a_i - r_i = a_{i+1} - r_{i+1}$ (or $a_k - r_k = 0$ if $i = k$).

In addition, there is a dual formula for these multi-Schur polynomials. For this, let

$$q_i = a_i - r_i \quad \text{for } 1 \leq i \leq k, \quad \text{and let}$$
$$n_1 = b_1 - r_1, \quad n_i = (b_i - r_i) - (b_{i-1} - r_{i-1}) \text{ for } 2 \leq i \leq k.$$

Then

$$\mathfrak{S}_w(x, y) = s_\mu^*((b_1, a_1)^{n_1}, \ldots, (b_k, a_k)^{n_k}),$$

where $\mu = (q_1{}^{n_1}, \ldots, q_k{}^{n_k})$, and s_μ^* is defined similarly to s_μ: it is the determinant of an $n \times n$ matrix, with $n = \sum n_i$, but using $e_l(u, v)$ in place of $h_l(u, v)$, where

$$e_l(u, v) = \text{ the coefficient of } T^l \text{ in } \prod_{i=1}^{u}(1 + x_i T) \Big/ \prod_{j=1}^{v}(1 + y_j T).$$

PROOF. This proposition is proved in [F2, §9.6] in the case where all the inequalities in (iii) and (iv) are strict, in which case the essential set of w is exactly the set $\{(a_1, b_1), \ldots, (a_k, b_k)\}$. The main point in that case is to construct the permutation w from the collections of integers, and to show that w is vexillary. Then it was known (cf. [M1]) that the double Schubert polynomials of vexillary permutations have determinantal expressions as in the proposition.

Here we show how to deduce this proposition from this irredundant case of strict inequalities. First, it is elementary to verify that (a_i, b_i) is not in the essential set if $b_i - r_i = b_{i-1} - r_{i-1}$ or $a_i - r_i = a_{i+1} - r_{i+1}$. The first case says that if the upper left $a_{i-1} \times b_{i-1}$ corner of a matrix has rank at most r_{i-1}, and $b_i - r_i = b_{i-1} - r_{i-1}$, then the upper left $a_i \times b_i$ corner has rank at most r_i. Indeed, the upper left $a_{i-1} \times b_i$ corner has rank at most $r_{i-1} + (b_i - b_{i-1}) = r_i$, so the smaller $a_i \times b_i$ corner also has rank at most r_i. The proof in the other case is similar. The fact that (a_1, b_1) or (a_k, b_k) is not essential if $r_1 = b_1$ or $r_k = b_k$ is obvious from the definition.

We prove the proposition by induction on the number of equalities occurring in equations (iii) or (iv) of the proposition. If there are such equalities, then some $p_i = p_{i+1}$, or $p_k = 0$, or some $m_i = 0$. If $m_i = 0$, we can simply omit the corresponding pair (a_j, b_j), for $j = k + 1 - i$, and apply the inductive case to the sequences of length $k - 1$, omitting the terms a_j, b_j and r_j, and omitting p_i and m_i. Note that the term $p_i{}^{m_i} = p_i{}^0$ does not appear in λ, so that the matrix whose determinant is the multi-Schur polynomial is unchanged by this omission.

We may therefore assume that all exponents m_i are positive. If $p_k = 0$, the last m_k rows of the matrix have 1's on the diagonal, with 0's to the left; therefore its determinant is unchanged by omitting the last m_k rows and columns. By induction, this determinant is that for the sequences of length $k - 1$ obtained by omitting a_1, b_1, and r_1, and omitting p_k and m_k.

The last case to consider is when some $p_i = p_{i+1}$, which means that $b_j - r_j = b_{j-1} - r_{j-1}$, for $j = k + 1 - i$. We omit a_j, b_j and r_j, and apply induction to the sequences of length $k - 1$; although p_i is omitted, m_{i+1} gets replaced by $m_i + m_{i+1}$. The partition λ is therefore the same, but the matrix changes. The change occurs only in a sequence of m_i rows, where one matrix has entries $h_l(b_j, a_j)$ and the other has entries $h_l(b_{j-1}, a_{j-1})$, with the subscripts l determined by λ. These rows are followed by a sequence of m_{i+1} rows, where both matrices have the same entries $h_l(b_{j-1}, a_{j-1})$. We have the following basic relations, which follow immediately from the definitions of the polynomials $h_l(u, v)$:

$$h_l(b_j, a_j) = h_l(b_{j-1}, a_{j-1}) - g_1 h_{l-1}(b_{j-1}, a_{j-1}) +$$
(*)
$$\cdots + (-1)^c g_c h_{l-c}(b_{j-1}, a_{j-1}),$$

where g_s is the sth elementary symmetric polynomial in the variables y_β and x_γ for $b_{j-1} < \beta \le b_j$ and $a_j < \gamma \le a_{j-1}$, and $c = b_j - b_{j-1} + a_{j-1} - a_j$.

Since these rows under discussion all correspond to parts of the partition of size $p_i = p_{i+1}$, each entry directly under an h_l is an h_{l+1}. One can therefore transform one matrix into the other by a sequence of elementary row operations, using (*), to prove that their determinants are equal. First one adds a multiple of the top row below the rows where the matrices differ to the last row where they differ; then add multiples of these two rows to the one above, and so on (cf. [M1, (3.6)]). This completes the proof of the proposition. The dual case with the dual partition μ is entirely similar.

BOTT-SAMELSON SCHEMES

In Chapters 2 and 6 formulas for degeneracy loci were proved by induction on their dimensions, using a basic $\mathbb{P}^1$-bundle correspondence to go from one such locus to another. Here we show that an iteration of this idea constructs resolutions of singularities of these loci, and provides explicit Gysin formulas. Historically, when the base is a point, these Bott-Samelson varieties, and the corresponding Gysin maps, were used to prove formulas for Schubert varieties. We state the results for cohomology, but they are equally valid for Chow groups.

We start with the (A_{n-1}) case. We are given a vector bundle V of rank n on a scheme X, together with a complete flag of subbundles $V_{\bullet} : V_1 \subset \cdots \subset V_n = V$. Let $\mathcal{F}$ be the bundle of complete flags in V. For $w \in S_n$, we have the locus $Y_w \subset \mathcal{F}$ of flags $L_{\bullet}$ given by the conditions

$$\dim(L_p \cap V_q) \geq \#\{i \leq p : w(i) > n - q\} \text{ for } 1 \leq p, q \leq n - 1.$$

For any sequence $i_1, \ldots, i_l$ of integers between 1 and $n-1$, define the subvariety $\mathcal{Z}(i_1, \ldots, i_l)$ of $\mathcal{F} \times_X \mathcal{F} \times \cdots \times_X \mathcal{F}$ $(l+1$ factors) by

$$\mathcal{Z}(i_1, \ldots, i_l) = \left\{ (L_{\bullet}^{(0)}, L_{\bullet}^{(1)}, \ldots, L_{\bullet}^{(l)}) : L_j^{(p)} = L_j^{(p-1)} \ \forall \ j \neq i_p, \ 1 \leq p \leq l \right\}.$$

Let $p_j : \mathcal{Z}(i_1, \ldots, i_l) \to \mathcal{F}$ be the projection onto the jth factor, for $0 \leq j \leq l$. When $l = 1$, $\mathcal{Z}(i) = \mathcal{F} \times_{\mathcal{F}(i)} \mathcal{F}$ is the scheme used in the proofs in Chapter 2. For $l > 1$, there is a canonical isomorphism

$$\mathcal{Z}(i_1, \ldots, i_l) \cong \mathcal{Z}(i_1) \times_{\mathcal{F}} \mathcal{Z}(i_2) \times_{\mathcal{F}} \cdots \times_{\mathcal{F}} \mathcal{Z}(i_l).$$

These schemes fit into a diagram in which all maps are projections of $\mathbb{P}^1$-bundles, and all squares are fiber squares:

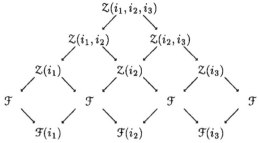

Let $E_1 \subset \cdots \subset E_n = V$ be the tautological flag on $\mathcal{F}$, and let $x_i = -c_1(E_i/E_{i-1})$. Let $p = p_0$ and $q = p_l$ be the two extreme projections. Define ∂_i as in Chapter 2.

PROPOSITION.

(1) $q_* \circ p^* : H^i(\mathcal{F}) \to H^{i-2l}(\mathcal{F})$ *is given by the formula*

$$q_* \circ p^* = \partial_{i_l} \circ \partial_{i_{l-1}} \circ \cdots \circ \partial_{i_1}.$$

(2) *For* $v \in S_n$, *let* $w = v \cdot s_{i_1} \cdot s_{i_2} \cdots \cdot s_{i_l}$. *If* $l(w) = l(v) - l$, *then* q *maps* $p^{-1}(Y_v)$ *birationally onto* Y_w.

PROOF. When $l = 1$, this is exactly the lemma from Chapter 2. The general case follows immediately by induction on l, using the above diagram and the following elementary facts about fiber squares of $\mathbb{P}^1$-bundles. More generally, given a fiber square

where r and s are smooth and projective morphisms, then

(i) $q_* \circ p^* = s^* \circ r_* : H^i(P) \to H^{i-2m}(Q)$, where m is the relative dimension of P over Y.

(ii) If r maps a subvariety V of P birationally onto a subvariety W of Y, then q maps $p^{-1}(V)$ birationally onto $s^{-1}(W)$.

The first of these is a special case of a general fact ([F1, §1.7]), and the second follows immediately from the definitions. □

Regarding $H^*(\mathcal{F})$ as a subalgebra of $H^*(\mathcal{Z}(i_1,\ldots,i_l))$ by means of q^*, we see that $H^*(\mathcal{Z}(i_1,\ldots,i_l))$ has generators $x_1{}^{m_1} \cdot x_2{}^{m_2} \cdot \ldots \cdot x_n{}^{m_n}$, where m_i is at most the number of times i occurs in the sequence $i_1,\ldots,i_l$. (This is verified by induction on l.) The proposition therefore determines the Gysin maps $q_* : H^i(\mathcal{Z}(i_1,\ldots,i_l)) \to H^{i-2l}(\mathcal{F})$:

(C.1) $q_*(x_1{}^{m_1} \cdot \ldots \cdot x_n{}^{m_n}) = \partial_{i_l} \circ \cdots \circ \partial_{i_1}(x_1{}^{m_1} \cdot \ldots \cdot x_n{}^{m_n}).$

It follows that $q_*(P) = \partial_{i_l} \circ \cdots \circ \partial_{i_1}(P)$ for any polynomial P in variables $x_1,\ldots,x_n$.

The most important special case is when $v = w_0$, in which case $Y_{w_0} = X \subset \mathcal{F}$ is the locus where $L_\bullet = V_\bullet$. In this case, if

$$w = w_0 \cdot s_{i_1} \cdot s_{i_2} \cdot \ldots \cdot s_{i_l}, \qquad \text{with } l(w) = l(w_0) - l,$$

then $\tilde{Y}_w = p^{-1}(Y_{w_0})$ is a subscheme of $\mathcal{Z}(i_1,\ldots,i_l)$ that is smooth over X (via p), and that maps birationally onto Y_w (via q). This $\tilde{Y}_w$ is called the **Bott-Samelson resolution** of Y_w. It is a subscheme of $\mathcal{F} \times_X \cdots \times_X \mathcal{F}$ (l factors):

$$\tilde{Y}_w = \big\{ \, (L_\bullet^{(1)},\ldots,L_\bullet^{(l)}) : L_j^{(1)} = V_j \text{ for } j \neq i_1,$$
$$\text{and } L_j^{(p)} = L_j^{(p-1)} \ \forall \, j \neq i_p, 2 \leq p \leq l \, \big\}.$$

The projection q maps $(L_\bullet^{(1)}, \ldots, L_\bullet^{(l)})$ to $L_\bullet^{(l)}$. This resolution depends on the choice of the sequence $s_{i_1}, \ldots, s_{i_l}$. We have

$$
\begin{array}{ccc}
\widetilde{Y}_w & \hookrightarrow & \mathcal{Z}(i_1, \ldots, i_l) \\
\downarrow & & \downarrow q \\
Y_w & \hookrightarrow & \mathcal{F}.
\end{array}
$$

In the particular case where $v = w_0$ and $w = e$, so $w_0 = s_{i_1} \cdot s_{i_2} \cdot \ldots \cdot s_{i_l}$ is a reduced word for w_0, then $Y_e = \mathcal{F}$, and $\widetilde{Y}_e \to \mathcal{F}$ is a birational morphism between smooth schemes over X.

Nothing important needs to be changed for the other classical cases. One starts with a vector bundle V on X with a symplectic or orthogonal form, and a fixed isotropic flag $V_1 \subset \cdots \subset V_n \subset V$; $\mathcal{F}$ is the bundle of isotropic flags $L_\bullet$ in V. The schemes $\mathcal{Z}(i_1, \ldots, i_l)$ are defined as for the (A_n) case. The only difference comes from the fact that in the orthogonal cases, the projections $\mathcal{F} \to \mathcal{F}(n)$ are not $\mathbb{P}^1$-bundles, but they are smooth conic bundles, which are almost as simple.

The scheme $\mathcal{Z}(i) \subset \mathcal{F} \times_X \mathcal{F}$ can be regarded as the graph of a correspondence from $\mathcal{F}$ to $\mathcal{F}$. In this language,

$$
\partial_i(\alpha) = q_* \circ p^*(\alpha) = \tilde{q}_*(\tilde{p}^*(\alpha) \cdot [\mathcal{Z}(i)]),
$$

where $\tilde{p}$ and $\tilde{q}$ are the projections from $\mathcal{F} \times_X \mathcal{F}$ to $\mathcal{F}$, and the dot denotes the intersection product. In general, $\mathcal{Z}(i_1, \ldots, i_l)$ is the variety used in constructing the product of the correspondences $\mathcal{Z}(i_1), \ldots, \mathcal{Z}(i_l)$, as in [F1, §16].

For a description from a group-theoretic point of view, see [D2] and [J]. For more about related geometry, see recent papers of Magyar, e.g. [Mag].

PFAFFIANS

Since handy references concerning Pfaffians are not readily available, we collect in this appendix some of their most useful properties.

Recall that if $X = (x_{ij})$ is a skew-symmetric matrix (i.e. $x_{ij} = -x_{ji}$ and $x_{ii} = 0$) of odd size, then its determinant is zero. On the other hand, if $X = (x_{ij})$ is a skew-symmetric matrix of even size $2n \times 2n$, then its determinant is a perfect square:

$$\det(X) = \mathrm{Pf}(X)^2,$$

where

$$\mathrm{Pf}(X) = \sum \mathrm{sgn}(\sigma)\, x_{\sigma(1)\sigma(2)} \cdot \ldots \cdot x_{\sigma(2n-1)\sigma(2n)}$$

summed over $\sigma \in S_{2n}$ such that $\sigma(2r - 1) < \sigma(2r)$ for $1 \le r \le n$, and $\sigma(2r - 1) < \sigma(2r + 1)$ for $1 \le r \le n - 1$. There are $(2n - 1) \cdot (2n - 3) \cdot \ldots \cdot 3 \cdot 1$ terms in this sum. Equivalently,

$$\mathrm{Pf}(X) = \frac{1}{2^n n!} \sum_{\sigma \in S_{2n}} \mathrm{sgn}(\sigma)\, x_{\sigma(1)\sigma(2)} \cdot \ldots \cdot x_{\sigma(2n-1)\sigma(2n)}.$$

EXAMPLE.

$$\mathrm{Pf} \begin{pmatrix} 0 & x_{12} \\ -x_{12} & 0 \end{pmatrix} = x_{12};$$

$$\mathrm{Pf} \begin{pmatrix} 0 & x_{12} & x_{13} & x_{14} \\ -x_{12} & 0 & x_{23} & x_{24} \\ -x_{13} & -x_{23} & 0 & x_{34} \\ -x_{14} & -x_{24} & -x_{34} & 0 \end{pmatrix} = x_{12}x_{34} - x_{13}x_{24} + x_{14}x_{23};$$

The Pfaffian of a 6×6 skew-symmetric matrix $X = (x_{ij})$ is

$$x_{12}x_{34}x_{56} - x_{12}x_{35}x_{46} \pm 13 \text{ similar terms.}$$

More generally, every minor of a skew-symmetric matrix X can be presented (in a non-unique way) as a quadratic expression in Pfaffians of skew-symmetric matrices obtained from X by removing certain rows and columns with the same numbers. For more on this, we refer the reader to [Hey, §3] and [B-E, Lemma 2.3 (2)].

A useful tool to compute with Pfaffians is Laplace-type expansion for them. Given a skew-symmetric matrix X of even size $2n \times 2n$, let $\mathrm{Pf}^{ij}(X)$ denote the Pfaffian of the skew-symmetric matrix obtained from X by removing the ith and j^{th} row and column. Then for a fixed integer j, $1 \leq j \leq 2n$, one has

$$(\mathrm{D}.1) \qquad \mathrm{Pf}(X) = \sum_{i<j}(-1)^{i+j-1}x_{ij}\,\mathrm{Pf}^{ij}(X) + \sum_{i>j}(-1)^{i+j}x_{ij}\,\mathrm{Pf}^{ij}(X).$$

This simple formula admits many generalizations presenting $\mathrm{Pf}(X)$ as quadratic expressions in subpfaffians of X, or as quadratic expressions in subpfaffians of X and minors of X (by a "subpfaffian of X" we understand the Pfaffian of the skew-symmetric matrix obtained from X by removing certain rows and columns with the same numbers). For more on this, see [Sr, §3] and [Ku, §5]. If we allow the rationals as the coefficients, then it is pretty easy to write down such an expression. Given a skew-symmetric matrix X of even size $2n \times 2n$, and a sequence of integers $1 \leq i_1 < \ldots < i_k \leq 2n$, let $\mathrm{Pf}_{i_1,\ldots,i_k}(X)$ denote the Pfaffian of the $k \times k$ skew-symmetric matrix formed by the entries of X, which belong to the rows and columns with numbers $i_1,\ldots,i_k$. Then for any $1 \leq m < n$, the following equality holds

$$(\mathrm{D}.2) \qquad \binom{n}{m}\mathrm{Pf}(X) = \sum_{\sigma}{}' \mathrm{sgn}(\sigma)\,\mathrm{Pf}_{\sigma(1),\ldots,\sigma(2m)}(X) \cdot \mathrm{Pf}_{\sigma(2m+1),\ldots,\sigma(2n)}(X),$$

where the summation $\sum'$ is over $\sigma \in S_{2n}$ such that $\sigma(1) < \ldots < \sigma(2m)$ and $\sigma(2m+1) < \ldots < \sigma(2n)$.

If X is a $2n \times 2n$ skew-symmetric matrix, then for any $2n \times 2n$ matrix A, the matrix AXA^T is skew-symmetric (A^T denotes the transposed matrix), and one has

$$(\mathrm{D}.3) \qquad\qquad \mathrm{Pf}(AXA^T) = \det(A) \cdot \mathrm{Pf}(X).$$

For the interesting history of Pfaffians, we recommend Knuth's article [Kn]. According to Knuth the following identity is basic. Before we state it, we introduce some notation. If $i_1,\ldots,i_k$ are different positive integers smaller than or equal to $2n$, but appearing not necessary in an ascending order, set

$$\mathrm{Pf}_{i_1,\ldots,i_k}(X) := \mathrm{sgn}(\sigma)\,\mathrm{Pf}_{i_{\sigma(1)}<\ldots<i_{\sigma(k)}}(X)$$

(where $\sigma \in S_k$). If $i_p = i_q$ for some $p \neq q$, then we put $\mathrm{Pf}_{i_1,\ldots,i_k}(X) = 0$. The identity reads as follows. For two sequences $1 \leq i_1 < \ldots < i_k \leq 2n$ and $1 \leq j_1 < \ldots < j_l \leq 2n$ where k and l are even,

$$\mathrm{Pf}_{i_1,\ldots,i_k}(X) \cdot \mathrm{Pf}_{i_1,\ldots,i_k,j_1,\ldots,j_l}(X)$$

$$(\mathrm{D}.4)$$
$$= \sum_{p=2}^{l}(-1)^p \mathrm{Pf}_{i_1,\ldots,i_k,j_1,j_p}(X) \cdot \mathrm{Pf}_{i_1,\ldots,i_k,j_2,\ldots,\widehat{j_p},\ldots,j_l}(X).$$

If the set of the i's is empty, then (D.4) specializes to (D.1). We refer to [Kn, §2 and 3] for variations of this identity and for its applications to the theory of Pfaffians and also determinants. In fact, as is said in [Kn, §3]:

> "Determinants are the special case of Pfaffians in which the index set is bipartite with respect to f, in the sense that $f[xy] = 0$ when x and y belong to the same part."

Here, $f = f[\bullet\bullet]$ means a skew-symmetric function (of two arguments) and the Pfaffian in question is $\mathrm{Pf}(f[x_i x_j])_{1 \le i,j \le 2n}$. Using this setup, it is worth mentioning the following property which is deduced in [Kn] from a special case of (D.4). The identity

$$(\text{D.5}) \qquad \mathrm{Pf}(f[x_i x_j])_{1 \le i,j \le 2n} = \prod_{1 \le i < j \le 2n} f[x_i x_j]$$

holds for all n if and only if it holds for $n = 2$ (see [Kn, p.158]).

For example, for

$$f[xy] = \frac{x - y}{x + y}$$

one verifies that

$$f[x_1 x_2] f[x_3 x_4] - f[x_1 x_3] f[x_2 x_4] + f[x_1 x_4] f[x_2 x_3]$$
$$= f[x_1 x_2] f[x_1 x_3] f[x_1 x_4] f[x_2 x_3] f[x_2 x_4] f[x_3 x_4].$$

By property (D.5) this implies that for any n,

$$(\text{D.6}) \qquad \mathrm{Pf}\left(\frac{x_i - x_j}{x_i + x_j}\right)_{1 \le i,j \le 2n} = \prod_{1 \le i < j \le 2n} \frac{x_i - x_j}{x_i + x_j}.$$

This last identity, playing an important role in Schur's original approach [Sch] to his Q-functions, was used in Chapter 8.

For some algebro-geometric purposes it is also important to consider the ideals generated by subpfaffians of a fixed order of (especially) a generic skew-symmetric matrix, i.e. the skew-symmetric matrix $X = (x_{ij})_{1 \le i,j \le m}$, where x_{ij} for $1 \le i < j \le m$ are independent variables. These ideals have nice algebraic properties; they are prime and Gorenstein, so in particular Cohen-Macaulay (see [Kl-La]), and serve as radicals of appropriate ideals generated by minors of a fixed order of X (see [B-E]). Their (characteristic zero) syzygies were determined in [J-P-W].

Geometrically, these ideals determine the scheme structures of some Schubert varieties in orthogonal Grassmannians. For more on this, see [Lk-Sh1].

GROUP-THEORETIC APPROACHES

For some of the problems treated in these lectures, it is natural to study them not "type by type" but uniformly for all algebraic groups at the same time. We sketch some of these results in this appendix. None of them is required for reading the text, but we hope they will be illuminating for those familiar with algebraic groups. A general reference for group-theoretic notions used in this appendix is [Hu].

Let G be a semisimple algebraic group and $B \subset G$ a Borel subgroup.

Let X be a variety on which B acts freely (from the right). Suppose that the quotient X/B exists so that $p : X \to X/B$ is a principal B-bundle. On the other hand, let $\mu : B \to GL(V)$ be a linear representation. We denote by $\mathcal{L}_\mu$ the vector bundle $X \times^B V$ that is the quotient of $X \times V$ by the equivalence relation

$$(x, v) \sim \left(xb, \mu(b)^{-1}v\right),$$

where $x \in X$, $b \in B$ and $v \in V$. Equivalently, if U is an open subset of X/B, then $\Gamma(U, \mathcal{L}_\mu)$ is the set of morphisms $\varphi : p^{-1}(U) \to V$ such that $\varphi(xb) = \mu(b)^{-1}\varphi(x)$.

In particular, with any character χ of B (that is, a homomorphism of B into the multiplicative group) there is associated a line bundle $\mathcal{L}_\chi$; this induces a homomorphism of groups $X(B) \to \mathrm{Pic}(X/B)$, where $X(B)$ denotes the group of characters of B.

Composing this homomorphism with the homomorphism of the first Chern class from $\mathrm{Pic}(X/B)$ to $H^2(X/B)$, one gets a homomorphism from $X(B)$ to $H^2(X/B)$, which extends to a homomorphism of graded rings

$$c : S^\bullet\left(X(B)\right) \to H^\bullet(X/B)$$

from the symmetric algebra of the $\mathbb{Z}$-module $X(B)$ to the cohomology ring of X/B; this homomorphism is called the **characteristic map** of the fiber bundle $p : X \to X/B$. In this appendix, $S = \oplus S^k$ will denote the symmetric algebra $S^\bullet\left(X(B)\right) = \oplus S^k\left(X(B)\right)$.

Choose a maximal torus $T \subset B$ with the Weyl group $W = N_G(T)/T$ of (G, T). Then W acts on the group of characters $X(T)$ of T, and since $X(B) = X(T)$, this induces an action of W on S. This action will be of a particular importance in this appendix.

The root system of (G, T) is denoted by R; the set R^+ of positive roots consists of the opposite elements of roots of (B, T). Let $\Delta \subset R^+$ be the associated basis of R. Denote by ρ half the sum of positive roots and by N their number. The

Weyl group W is generated by simple reflections, i.e. by the reflections associated with the elements of Δ. For any root $\alpha \in R$, we denote by s_α the reflection associated with α. (The reflection s_α can be realized as a linear endomorphism of the Euclidean space $X(T) \otimes \mathbb{R}$; the formula is: $s_\alpha(x) = x - \frac{2(x,\alpha)}{(\alpha,\alpha)}\alpha$.)

By a **reduced decomposition** of an element $w \in W$ we understand a presentation $w = s_{\alpha_1} \cdot \ldots \cdot s_{\alpha_l}$ where all $\alpha_p \in \Delta$, and l is the smallest number occuring in such a presentation, called the **length** of w and denoted $l(w)$.

By w_0 we denote the longest element of W, the unique element of W with length equal to N.

E.1 Gysin maps for complete flag bundles

Let $\pi : X \to Y$ be a principal G-bundle, where X and Y are nonsingular. Then π factors through the complete flag bundle $f : X/B \to Y$. The morphism f is smooth and proper and of relative dimension N.

Then for any $u \in S$, we have in $H^*(X/B)$,

$$(E.1) \qquad f^* f_* c(u) = c\left(\frac{\sum_{w \in W} \det(w)w(u)}{\prod_{\alpha \in R^+} \alpha}\right).$$

We sketch the proof of this identity, due to Brion [Br]. Since, for $\mu \in X(B)$, $f^* f_* \mathcal{L}_\mu$ is the vector bundle over X/B associated with the B-module $\Gamma(G/B, \mathcal{L}_\mu)$, the Chern roots of $f^* f_* \mathcal{L}_\mu$ are the images by c of the weights of $\Gamma(G/B, \mathcal{L}_\mu)$. Therefore for a dominant weight λ, the Weyl character formula implies

$$ch(f^* f_* \mathcal{L}_\lambda) = c\left(\frac{\sum_{w \in W} \det(w)e^{w(\lambda+\rho)}}{\prod_{\alpha \in R^+}(e^{\alpha/2} - e^{-\alpha/2})}\right),$$

where e^μ, for $\mu \in X(B)$, denotes $\sum_{n=0}^\infty \mu^n/n!$ (and thus $c(e^\mu)$ makes sense in $H^*(X/B)$ because $c(\mu)$ is nilpotent). Invoking the Grothendieck-Riemann-Roch theorem:

$$ch(f_* \mathcal{L}_\lambda) = f_*\big(ch(\mathcal{L}_\lambda) \cdot td(T_f)\big),$$

we thus get

$$f^* f_* c\left(e^\lambda \prod_{\alpha \in R^+} \frac{\alpha}{1 - e^{-\alpha}}\right) = c\left(\frac{\sum_{w \in W} \det(w)e^{w(\lambda+\rho)}}{\prod_{\alpha \in R^+}(e^{\alpha/2} - e^{-\alpha/2})}\right)$$

(note that the Chern roots of T_f, the relative tangent bundle, are $c(\alpha), \alpha \in R^+$). If now we set

$$u_0 := \prod_{\alpha \in R^+} \frac{\alpha}{e^{\alpha/2} - e^{-\alpha/2}}$$

and $\mu := \lambda + \rho$, then we get the assertion for $u = u_0 e^\mu$, with μ dominant and regular:

$$f^* f_*(u_0 e^\mu) = c\left(u_0 \frac{\sum_{w \in W} \det(w)e^{w(\mu)}}{\prod_{\alpha \in R^+} \alpha}\right) = c\left(\frac{\sum_{w \in W} \det(w)w(u_0 e^\mu)}{\prod_{\alpha \in R^+} \alpha}\right)$$

(observe that u_0 is W-invariant). As shown in [Br], this implies the identity for any $u \in S$.

Formula (E.1) offers a uniform generalization to all semisimple algebraic groups of some formulas for Gysin maps for flag bundles from Chapters 4 and 7. It is possible to generalize it further by replacing a Borel group B by any parabolic subgroup $G \supset P \supset B$ (i.e. geometrically by replacing complete flag bundles by partial flag bundles). For more about this, we refer the reader to [Br]. There the reader will also find a formula for the Todd class of the relative tangent bundle of such flag bundles.

E.2 Schubert varieties and the cohomology rings of G/P

In the geometry of flag manifolds G/B a large role is played by the Schubert cells BwB/B and their closures, the Schubert varieties $\overline{BwB/B}$. We set $X_w = \overline{[BwB/B]}$ in $H^*(G/B)$, and set $Y_w = X_{w_0 w} \in H^{2l(w)}(G/B)$. The Schubert cells form a cellular decomposition of G/B, so the classes X_w or Y_w form an additive basis for the cohomology.

The characteristic map

$$c : S \to H^*(G/B)$$

of the fibration $G \to G/B$ is usually called the **Borel characteristic map**. Its kernel is generated by positive degree W-invariants, and $c \otimes \mathbb{Q}$ is surjective.

Suppose $P \in S^k$. The key problem is to understand the expansion of $c(P)$ as a linear combination of Schubert classes. This is closely related to the problem of finding a polynomial P so that $c(P)$ is a given Schubert class. As we have seen in the text, the general answer involves divided difference operators.

Here is the general description of these operators. For any root $\alpha \in R$ and $u \in S$, α divides $u - s_\alpha(u)$, and thus the formula

$$\partial_\alpha(u) = \frac{u - s_\alpha(u)}{\alpha}$$

defines a well-defined operator $\partial_\alpha : S \to S$, lowering degree by 1. The reader finds in [B-G-G] and [D1] two different proofs of the following basic fact. Let $w = s_{\alpha_1} \cdot \ldots \cdot s_{\alpha_l} = s_{\beta_1} \cdot \ldots \cdot s_{\beta_l}$ be two reduced decompositions of w. Then

$$\partial_{\alpha_1} \circ \ldots \circ \partial_{\alpha_l} = \partial_{\beta_1} \circ \ldots \circ \partial_{\beta_l}$$

and consequently there is a well-defined operator $\partial_w : S \to S$ not depending on the reduced decomposition chosen.

Coming back to our problem, the answer is:

$$(E.2) \qquad\qquad c(P) = \sum_{l(w)=k} \partial_w(P) Y_w.$$

There is an action of ∂_w on $H^*(G/B)$ that is compatible with c. In fact, define $\partial_w : H^*(G/B) \to H^*(G/B)$ by the formula

$$(E.3) \qquad\qquad \partial_w(Y_v) = \begin{cases} Y_{vw^{-1}} & \text{if } l(vw^{-1}) = l(v) - l(w) \\ 0 & \text{otherwise.} \end{cases}$$

The fact that $\partial_w \circ c = c \circ \partial_w$ can be proved using $\mathbb{P}^1$-correspondences as in Chapters 2 and 6.

In particular, we have

$$(E.4) \qquad\qquad Y_w = \partial_{w^{-1}w_0}(Y_{w_0}).$$

The top class Y_{w_0} can be expressed using the characteristic map by the following two expressions, with rational coefficients

$$(E.5) \qquad c\left(\frac{1}{|W|}\prod_{\alpha \in R^+}\alpha\right) = Y_{w_0} \qquad \text{and} \qquad c\left(\frac{\rho^N}{N!}\right) = Y_{w_0}$$

(see [B-G-G]).

A similar theory works for parabolic subgroups. Let $P_\theta \supset B$ be the parabolic subgroup associated with a subset $\theta \subset \Delta$ (see [Hu, §30]). Denote by W^θ the set

$$W^\theta := \{w \in W : l(ws_\alpha) = l(w) + 1 \quad \forall \alpha \in \theta\}.$$

This last set is the set of minimal length left coset representatives of W_θ, the subgroup of W generated by $\{s_\alpha\}_{\alpha \in \theta}$, in W. The projection $G/B \to G/P_\theta$ induces an inclusion $H^*(G/P_\theta) \hookrightarrow H^*(G/B)$ which additively identifies $H^*(G/P_\theta)$ with $\bigoplus_{w \in W^\theta} \mathbb{Z}Y_w$. Multiplicatively, $H^*(G/P_\theta)$ is identified with the ring of invariants $H^*(G/B)^{W_\theta}$. The restriction

$$c : S^{W_\theta} \to H^*(G/P_\theta)$$

of the Borel characteristic map satisfies, for any W_θ-invariant P from S^k, the equation

$$(E.6) \qquad\qquad c(P) = \sum_{\substack{w \in W^\theta \\ l(w)=k}} \partial_w(P)Y_w.$$

This equation allows one at least in theory to compute the Schubert-class-expansion of the product of any Schubert class by any element of $H^*(G/P_\theta)$, following a strategy that we describe now. Our goal is to compute the coefficients m_w^v appearing in

$$Y_w \cdot c(Q) = \sum_v m_w^v Y_v,$$

where $w, v \in W^\theta$ and $Q \in S^l$ is W_θ-invariant. To achieve this goal, one uses the Leibniz-type formula

$$\partial_\alpha(P \cdot Q) = P \cdot \partial_\alpha(Q) + \partial_\alpha(P) \cdot s_\alpha(Q).$$

Let $P \in S^k$ be W_θ-invariant and such that $c(P) = Y_w$. We have, for a reduced decomposition $v = s_{\alpha_1} \cdot \ldots \cdot s_{\alpha_{k+l}}$,

$$m_w^v = \partial_v(P \cdot Q) = \partial_{\alpha_1} \circ \ldots \circ \partial_{\alpha_{k+l}}(P \cdot Q)$$
$$= \sum \partial_J(P)\partial_\alpha^J(Q),$$

where the sum is over all subsequences $J = (j_1 < \ldots < j_k) \subset \{1, 2, \ldots, k+l\}$; ∂_J is ∂_{r_J} where $r_J = s_{\alpha_{j_1}} \cdot \ldots \cdot s_{\alpha_{j_k}}$, and ∂_α^J is obtained by replacing in $\partial_{\alpha_1} \circ \ldots \circ \partial_{\alpha_{k+l}}$ each ∂_{α_j} by s_{α_j} for $j \in J$. By the choice of P, we get

$$m_w^v = \sum \partial_\alpha^J(Q),$$

where the sum runs over all J such that r_J is a reduced decomposition of w.

This strategy was applied, in a series of papers [P-R2-4], to establish Pieri-type formulas for the spaces G/P_θ, where G is a classical group and θ stems from Δ by omitting a single simple root α. This is technically too involved to be discussed here, so we refer the interested reader to the original papers or to the summary of this theory in [P4, §6]. At least in theory one can also try to understand the multiplication of Schubert classes via the following Chevalley's formula [Ch]: in $H^*(G/B)$ one has for $w \in W$ and a simple root α,

$$Y_w \cdot Y_{s_\alpha} = \sum (\beta^\vee, \omega_\alpha) Y_{w s_\beta},$$

where β runs over positive roots such that $l(w s_\beta) = l(w) + 1$ and ω_α denotes the fundamental weight associated with the simple root α. In the (A_n) case, this formula was established independently by Monk (see Chapter 2) and has played an important role in the development of the Lascoux-Schützenberger theory of Schubert polynomials. (By the way, the Chevalley formula can be easily proven using the above strategy.) In practice, however, divided difference computations are better suited to this problem than the Chevalley formula.

The characteristic map is also useful to study the cohomology of other varieties with a B-action. For example, the cohomology of Bott-Samelson schemes in [D2] is studied in this way.

E.3 The classes of diagonals in flag bundles

We end this appendix with a certain characterization of the class of the relative diagonal in a flag bundle.

Let $\pi : X \to Y$ be a principal G-bundle where X and Y are nonsingular. Then π factors through the complete flag bundle $f : X/B \to Y$. We have the characteristic map

$$c : S = S^\bullet\big(X(B)\big) \to H^*(X/B)$$

and denote by $c' : S \otimes_{\mathbf{Z}} S \to H^*(X/B \times_Y X/B)$ the composition

$$S \otimes_{\mathbf{Z}} S \xrightarrow{c \otimes c} H^*(X/B) \otimes_{\mathbf{Z}} H^*(X/B) \to H^*(X/B \times_Y X/B).$$

The last map is defined by taking $a \otimes b$ to $p_1^*(a) \cdot p_2^*(b)$, where $p_i : X/B \times_Y X/B \to X/B$ denote the two projections. The following criterion, due to Graham [Gra], says when $P = P(x, y) \in S \otimes_{\mathbf{Z}} S$ represents the class of the relative diagonal $\Delta \subset X/B \times_Y X/B$.

CRITERION. For $P \in S \otimes_{\mathbf{Z}} S$, c' maps P onto the class of diagonal $\Delta \subset X/B \times_Y X/B$ if and only if the two properties hold:

(1) (Vanishing property) For every $w \in W$, $w \neq id$, one has $P(x, wx) = 0$.
(2) (Normalization property) $m\big(P(x,x)\big) = \prod_{\alpha \in R^+} \alpha \in S$, where m denotes multiplication $m : S \otimes_{\mathbf{Z}} S \to S$.

For a proof and some applications of this criterion, we refer the reader to [Gra] (where a somewhat different setup is used).

In the (A_n) case this characterization was established also by Lascoux [Las2] to characterize the double Schubert polynomials.

It is pointed out at the end of the introduction to [Gra] that this criterion can also be obtained using equivariant cohomology and some results of Arabia and Kostant-Kumar quoted therein.

De Concini (private communication) discovered independently (but later) this criterion using equivariant cohomology. His approach is also constructive, i.e. it produces P satisfying (1) and (2) in an explicit way. For instance, one of the expressions he obtains in the (C_n) case is

$$\prod_{1 \le i < j \le n} (y_i^2 - x_j^2) \prod_{1 \le i \le n} (y_i + x_i).$$

PUSH-FORWARD FORMULAS

The goal of this appendix is to provide a proof of the proposition from Section 4.1.[1]

We prove first (i). Let $q = 1$ and $\xi = c_1(Q)$. It follows from the identity

$$s_i(S - F_\mathcal{G}) = s_i(E_\mathcal{G} - F_\mathcal{G}) - \xi s_{i-1}(E_\mathcal{G} - F_\mathcal{G})$$

that for $\mu = (\mu_1, \dots, \mu_{n-1})$, where $n = \operatorname{rank}(E)$,

$$s_\mu(S - F_\mathcal{G}) = \begin{vmatrix} 1 & \xi & \cdots & \xi^{n-1} \\ s_{\mu_1-1}(E_\mathcal{G} - F_\mathcal{G}) & s_{\mu_1}(E_\mathcal{G} - F_\mathcal{G}) & \cdots & s_{\mu_1+n-2}(E_\mathcal{G} - F_\mathcal{G}) \\ s_{\mu_2-2}(E_\mathcal{G} - F_\mathcal{G}) & s_{\mu_2-1}(E_\mathcal{G} - F_\mathcal{G}) & \cdots & s_{\mu_2+n-3}(E_\mathcal{G} - F_\mathcal{G}) \\ \vdots & \vdots & \ddots & \vdots \end{vmatrix}.$$

For $k \geq 0$ and $l \geq 0$, we have

$$\pi_*\big(\xi^k s_l(Q - F_\mathcal{G})\big) =$$
$$= \pi_*\big(\xi^{k+l} - \xi^{k+l-1}c_1(F_\mathcal{G}) + \xi^{k+l-2}c_2(F_\mathcal{G}) - \dots + (-1)^l \xi^k c_l(F_\mathcal{G})\big).$$

If $k + l < n - 1$, then this expression is zero; if $k + l = n - 1$, it equals 1. Assuming additionally that $k \leq n - 1$ (which is our case), we get, for $k + l > n - 1$, that this expression is equal to

$$s_{k+l-n+1}(E) - s_{k+l-n}(E)c_1(F) + \dots + (-1)^{k+l-n+1}c_{k+l-n+1}(F) = s_{k+l-n+1}(E-F).$$

Hence we infer

$$\pi_*\big(s_l(Q - F_\mathcal{G}) \cdot s_{\mu_1,\dots,\mu_{n-1}}(S - F_\mathcal{G})\big)$$

$$= \begin{vmatrix} s_{l-n+1}(E - F) & s_{l-n+2}(E - F) & \cdots & s_l(E - F) \\ s_{\mu_1-1}(E - F) & s_{\mu_1}(E - F) & \cdots & s_{\mu_1+n-2}(E - F) \\ \vdots & \vdots & \ddots & \vdots \end{vmatrix}$$

$$= s_{l-n+1,\mu_1,\dots,\mu_{n-1}}(E - F).$$

---

[1] The formulas of this proposition were discussed in [P1] on pp.421-422. This appendix, in fact, corrects some defects and clarifies the exposition of that part of [P1].

To make the induction step from $q - 1$ to q, we use the following commutative diagram of Grassmann bundles and the corresponding tautological sequences:

(*)
$$G^{q-1}(\pi_1^* S') \cong Fl^{q,1} E \cong G^1(\pi_3^* Q)$$

$$0 \to S \to \pi_1^* S' \to P \to 0 \quad \pi_1 \qquad \pi_3 \quad 0 \to P \to \pi_3^* Q \to \mathcal{O}(1) \to 0$$

$$G^1 E \qquad\qquad G^q E$$

$$0 \to S' \to \pi_2^* E \to \mathcal{O}(1) \to 0 \quad \pi_2 \qquad \pi \quad 0 \to S \to \pi^* E \to Q \to 0$$

$$X$$

Here, $Fl^{q,1} E$ is the flag variety parametrizing flags of quotients of E of ranks q and 1. The diagram (*) gives two ways of constructing $Fl^{q,1} E$ as a Grassmann bundle.

We have (we omit writing the pull-back indices)

$$\pi_* \big(s_{\lambda_1, \ldots, \lambda_q}(Q - F) \cdot s_{\mu_1, \ldots, \mu_r}(S - F) \big)$$
$$= \pi_* (\pi_3)_* \big[s_{\lambda_1 + q - 1}(\mathcal{O}(1) - F) \cdot s_{\lambda_2, \ldots, \lambda_q}(P - F) \cdot s_{\mu_1, \ldots, \mu_r}(S - F) \big]$$
$$= (\pi_2)_* \big[s_{\lambda_1 + q - 1}(\mathcal{O}(1) - F) \cdot (\pi_1)_* \big(s_{\lambda_2, \ldots, \lambda_q}(P - F) \cdot s_{\mu_1, \ldots, \mu_r}(S - F) \big) \big]$$
$$= (\pi_2)_* \big[s_{\lambda_1 + q - 1}(\mathcal{O}(1) - F) \cdot s_{\lambda_2 - r, \ldots, \lambda_q - r, \mu_1, \ldots, \mu_r}(S' - F) \big]$$
$$= s_{\lambda_1 - r, \lambda_2 - r, \ldots, \lambda_q - r, \mu_1, \ldots, \mu_r}(E - F).$$

Here, the first and fourth equalities follow from the case $q = 1$, and the third equality follows from the induction assumption. This proves (i).

As for (ii), we also prove first the case $q = 1$. Assume that μ is a partition of length p which is arbitrary. We have

$$s_\mu(S - F_\mathcal{G}) = \begin{vmatrix} 1 & \xi & \cdots & \xi^p \\ s_{\mu_1 - 1}(E_\mathcal{G} - F_\mathcal{G}) & s_{\mu_1}(E_\mathcal{G} - F_\mathcal{G}) & \cdots & s_{\mu_1 + p - 1}(E_\mathcal{G} - F_\mathcal{G}) \\ s_{\mu_2 - 2}(E_\mathcal{G} - F_\mathcal{G}) & s_{\mu_2 - 1}(E_\mathcal{G} - F_\mathcal{G}) & \cdots & s_{\mu_2 + p - 2}(E_\mathcal{G} - F_\mathcal{G}) \\ \vdots & \vdots & \ddots & \vdots \end{vmatrix}.$$

Now, for $l \geq m = \mathrm{rank}(F)$ and $k \geq 0$, we have

$$\pi_* \big(\xi^k s_l(Q - F_\mathcal{G}) \big)$$
$$= \pi_* \big(\xi^{k+l} - \xi^{k+l-1} c_1(F_\mathcal{G}) + \xi^{k+l-2} c_2(F_\mathcal{G}) - \ldots + (-1)^m \xi^{k+l-m} c_m(F_\mathcal{G}) \big)$$
$$= s_{k+l-n+1}(E - F).$$

(Informally speaking, this holds because $l \geq m$ implies that $s_l(Q - F_{\mathcal{G}})$ "contains" all possible Chern classes of $F_{\mathcal{G}}$, and so we get the "complete" $s_{k+l-n+1}(E - F)$.) Therefore, we get

$$\pi_*\Big(s_l(Q - F_{\mathcal{G}}) \cdot s_\mu(S - F_{\mathcal{G}})\Big) = s_{l-n+1,\mu_1,\dots,\mu_p}(E - F).$$

Then the induction step from $q-1$ to q can be made using the diagram (*). Indeed, we have (we omit writing the pull-back indices)

$$
\begin{aligned}
\pi_*\big(s_{\lambda_1,\dots,\lambda_q}(Q - F) \cdot s_{\mu_1,\dots,\mu_p}(S - F)\big) & \\
= \pi_*(\pi_3)_*\big[s_{\lambda_1+q-1}(\mathcal{O}(1) - F) \cdot s_{\lambda_2,\dots,\lambda_q}(P - F) \cdot s_{\mu_1,\dots,\mu_p}(S - F)\big] & \\
= (\pi_2)_*\big[s_{\lambda_1+q-1}(\mathcal{O}(1) - F) \cdot (\pi_1)_*\big(s_{\lambda_2,\dots,\lambda_q}(P - F) \cdot s_{\mu_1,\dots,\mu_p}(S - F)\big)\big] & \\
= (\pi_2)_*\big[s_{\lambda_1+q-1}(\mathcal{O}(1) - F) \cdot s_{\lambda_2-r,\dots,\lambda_q-r,\mu_1,\dots,\mu_p}(S' - F)\big] & \\
= s_{\lambda_1-r,\lambda_2-r,\dots,\lambda_q-r,\mu_1,\dots,\mu_p}(E - F). &
\end{aligned}
$$

This proves (ii).

REMARK. The formula in (i) is valid, in fact, for all $\lambda \in \mathbb{Z}^q$ and $\mu \in \mathbb{Z}^r$. Of course, it contains nontrivial information only if $\lambda_i \geq -(q - i)$ for every i and $\mu_j \geq -(r - j)$ for every j, since otherwise it simply asserts that $0 = 0$.

EXERCISE. In the setup of (i), prove the following generalization of the formula stated in (i). Let $F_1, \dots, F_n$ be vector bundles on X. Then

$$\pi_*\Big(s_\lambda(Q - *) \cdot s_\mu(S - \square)\Big) = s_{\lambda_1-r,\dots,\lambda_q-r,\mu_1,\dots,\mu_r}(E - \bullet),$$

where the asterisk is replaced by $(F_i)_{\mathcal{G}}$ in the ith row of the Schur determinant for s_λ, the square is replaced by $(F_{q+i})_{\mathcal{G}}$ in the ith row of the Schur determinant for s_μ, and the dot is replaced by F_i in the ith row of the Schur determinant on the right-hand side.

THE CLASSES OF RELATIVE DIAGONALS

In this appendix, we discuss some useful results about the classes of relative diagonals, which stem essentially from [P4, §5] (see also [Gra]).

We are working simultaneously in topological and algebraic geometry categories, using singular cohomology rings and operational Chow rings respectively.

In topology, we assume that $\pi : \mathcal{G} \to X$ is the projection of a fiber bundle, with compact real oriented fibers. In algebraic geometry, we assume $\pi : \mathcal{G} \to X$ is a smooth proper morphism of algebraic schemes over a field. These assumptions guarantee that we have Gysin pushforward maps on cohomology rings, satisfying the projection formula $\pi_*(\pi^*(x) \cdot y) = x \cdot \pi_*(y)$. See [F1, §17] or [F-M] for this general formalism for singular spaces. In most applications, such as those in this text, the base is also smooth, in which case one can use Poincaré duality, identifying cohomology with homology, to define the Gysin maps in topology, and one can replace the operational Chow rings by the usual Chow groups in algebraic geometry.

To avoid awkward signs, we will make the simplifying assumption in topology that the fibers of π have even real dimension. Our crucial assumption is that $\pi^* : H^*(X) \to H^*(\mathcal{G})$ makes $H^*(\mathcal{G})$ a free *right* $H^*(X)$-module with a basis $\{a_\alpha\}$ and that π^* makes $H^*(\mathcal{G})$ a free *left* $H^*(X)$-module with a basis $\{b_\alpha\}$ (note that the indexing set for these two bases is the same).[1] Let δ denote the diagonal embedding of $\mathcal{G}$ in $\mathcal{G} \times_X \mathcal{G}$. We assume that we have a formula for the diagonal:

$$\delta_*(1) = \sum_{\alpha,\beta} p_1^*(a_\alpha) \cdot d_{\alpha\beta} \cdot p_2^*(b_\beta),$$

where $d_{\alpha\beta} \in H^*(X)$ and $p_i : \mathcal{G} \times_X \mathcal{G} \to \mathcal{G}$, $i = 1, 2$, are the projections onto the factors. (In this formula, as is customary, the cohomology class $d_{\alpha\beta}$ on X is identified with its pullback to $\mathcal{G} \times_X \mathcal{G}$.)

This last property holds, for instance, if π is a topologically locally trivial fiber bundle with fiber F and $\{a_\alpha\}$ and $\{b_\alpha\}$ are two families of elements of $H^*(\mathcal{G})$ which restrict to bases of $H^*(F)$. Indeed, by the Leray-Hirsch theorem, $H^*(\mathcal{G})$ is free over $H^*(X)$ with basis $\{a_\alpha\}$ or $\{b_\alpha\}$, and the natural map taking $a \otimes b$ to $p_1^*(a) \cdot p_2^*(b)$ establishes an isomorphism

$$H^*(\mathcal{G}) \otimes_{H^*(X)} H^*(\mathcal{G}) \xrightarrow{\sim} H^*(\mathcal{G} \times_X \mathcal{G}).$$

With these assumptions, we have the following basic fact.

---

[1] In topology, if some of these classes have odd degree, one must distinguish between left and right bases.

PROPOSITION. *For every $g \in H^*(\mathcal{G})$,*

$$g = \sum_{\alpha,\beta} a_\alpha \cdot d_{\alpha\beta} \cdot \pi_*(b_\beta \cdot g).$$

PROOF. Since $p_1 \circ \delta$ and $p_2 \circ \delta$ are the identity on $\mathcal{G}$, we have, for $g \in H^*(\mathcal{G})$,

$$
\begin{aligned}
g = p_{1*}\delta_*(g) &= p_{1*}\delta_*(\delta^* p_2^*(g)) \\
&= p_{1*}(p_2^*(g) \cdot \delta_*(1)) \\
&= p_{1*}(\delta_*(1) \cdot p_2^*(g)) \\
&= p_{1*}\left(\sum_{\alpha,\beta} p_1^*(a_\alpha) \cdot d_{\alpha\beta} \cdot p_2^*(b_\beta \cdot g)\right) \\
&= \sum_{\alpha,\beta} a_\alpha \cdot d_{\alpha\beta} \cdot \pi_*(b_\beta \cdot g).
\end{aligned}
$$

Here the fact that δ has even real codimension was used to commute $\delta_*(1)$ with $p_2^*(g)$, and the identity $p_{1*} \circ p_2^* = \pi^* \circ \pi_*$ (cf. [F1, Prop. 1.7 and §17.2]) associated with the fibre square

$$
\begin{array}{ccc}
\mathcal{G} \times_X \mathcal{G} & \xrightarrow{\ p_1\ } & \mathcal{G} \\
{\scriptstyle p_2}\Big\downarrow & & \Big\downarrow{\scriptstyle \pi} \\
\mathcal{G} & \xrightarrow{\ \pi\ } & X
\end{array}
$$

was used in the last step. □

Let us record two useful consequences of this proposition.

COROLLARY 1. *Denote by C the matrix $\left(\pi_*(b_\alpha \cdot a_\beta)\right)$ and by D the matrix $(d_{\alpha\beta})$. Then $D \cdot C = I$, the identity matrix.*

Indeed, applying the proposition to $g = a_\gamma$,

$$a_\gamma = \sum_{\alpha,\beta} a_\alpha \cdot d_{\alpha\beta} \cdot \pi_*(b_\beta \cdot a_\gamma),$$

and since $\{a_\alpha\}$ is a basis,

$$\sum_\beta d_{\alpha\beta}\, \pi_*(b_\beta \cdot a_\gamma) = \delta_{\alpha\gamma}$$

(the Kronecker delta), which gives the assertion.

COROLLARY 2. *With the above notation and assumptions, the following two conditions are equivalent:*
  (i) For any α, β, $d_{\alpha\beta} = \delta_{\alpha\beta}$.
  (ii) For any α, β, $\pi_*(b_\alpha \cdot a_\beta) = \delta_{\alpha\beta}$.

These corollaries strengthen, and the proofs simplify, results of [P4, §5] and [Gra], where the reader will find applications of the above corollaries to computation of the fundamental classes of subschemes, as well as to derivation of some expressions for the classes of relative diagonals in flag bundles.

A CONSTRUCTION OF MUMFORD

Assume that E is a rank n vector bundle on a smooth complete complex curve C (any algebraically closed field of characteristic different from 2 will also work). Suppose that E is endowed with a nondegenerate quadratic form $Q : E \otimes E \to \Omega_C$. Mumford's construction allows one to present $\Gamma(C, E)$ as the intersection of two maximal isotropic subspaces in a certain "big" vector space equipped with some nondegenerate quadratic form.

Fix a sufficiently positive divisor $D = \sum_{i=1}^{d} P_i$ where P_i are distinct points on C. Consider a commutative diagram of vector bundles on C

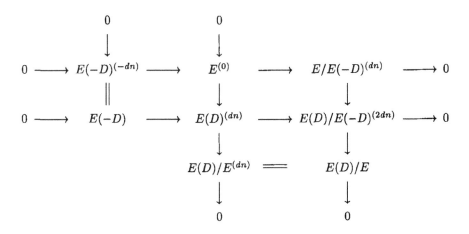

The superscripts in parentheses associated with the displayed bundles correspond to their Euler characteristics. Indeed, the existence of Q implies $E \cong \text{Hom}(E, \Omega_C)$, whence by the Serre duality, $\chi(E) = 0$. Consequently, $\chi(E(D)) = dn$, $\chi(E(-D)) = -dn$, and, by additivity, $\chi(E(D)/E(-D)) = 2dn$.

The above diagram induces a commutative diagram of vector spaces

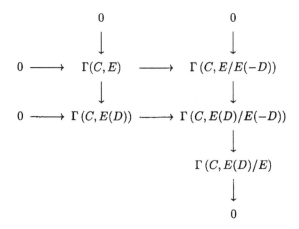

Set

$$V := \Gamma(C, E(D)/E(-D)), \quad W := \Gamma(C, E(D)), \quad U := \Gamma(C, E/E(-D)).$$

Since D is sufficiently positive, $H^1(C, E(D)) = 0$ and so $\dim(W) = dn$. Also $H^1(C, E(D)/E(-D)) = 0$, whence $\dim(V) = 2dn$. Since $H^1(C, E(D)/E) = 0$, we have $\dim(\Gamma(C, E(D)/E)) = dn$. The last two equalities imply $\dim(U) = dn$.

We now define a quadratic form $\Phi : V \otimes V \to \mathbb{C}$. For $v, w \in \Gamma(C, E(D)/E(-D))$, let v_i, w_i denote their stalks at P_i. Consider the shift $E(D) \otimes E(D) \to \Omega_C(2D)$ of Q, and denote it by the same letter. We set

$$\Phi(v \otimes w) := \sum_{i=1}^{d} \operatorname{Res}_{P_i}\big(Q(v_i \otimes w_i)\big).$$

The form Φ is clearly nondegenerate. Observe that if $v, w \in W$ then $Q(v \otimes w)$ is a globally defined rational form, and hence $\Phi(v \otimes w) = 0$ by the Residue Theorem (see, e.g., [A-C-G-H]). If $v, w \in U$ then $v_i, w_i \in (\Omega_C)_{P_i}$ and thus $\Phi(v \otimes w) = 0$. Therefore $W, U \subset V$ are maximal isotropic subspaces w.r.t. Φ. Finally, as it immediately follows from the above diagram,

$$\Gamma(C, E) = W \cap U \subset V.$$

This is the content of Mumford's construction [Mu1].

SCHUR-WEYL MODULES

There are fundamental relations among the geometry of flag manifolds, the combinatorics of Young diagrams and the symmetric groups, and representation theory. These ties to representation theory are not emphasized in the text, but it may be worthwhile to use this opportunity to give a simple construction of the Schur-Weyl modules, which can be constructed in a way similar to, and generalizing, the construction of symmetric and exterior powers that are standard in commutative algebra. For more about this, which comes from an idea of J. Towber, see [F6].

This construction makes sense for any module E over a commutative ring R. For any partition λ, it produces an R-module that can be denoted[1] $S^\lambda E$ or simply E^λ. The two extremes when $\lambda = (n)$ or $\lambda = (1^n)$ are the well-known symmetric and exterior powers $S^n E = \text{Sym}^n(E)$ and $\wedge^n E = \text{Alt}^n(E)$, constructed as solutions to universal problems. For example, there is a map from the Cartesian product $E^{\times n}$ to $\wedge^n E$ that takes a product $v_1 \times \cdots \times v_n$ to an element denoted $v_1 \wedge \ldots \wedge v_n$; this map is multilinear and alternating, and *universal*: for any R-module F and any map $\phi: E^{\times n} \to F$ that is multilinear and alternating, there is a unique homomorphism of R-modules $\tilde{\phi}: \wedge^n E \to F$ such that

$$\tilde{\phi}(v_1 \wedge \ldots \wedge v_n) = \phi(v_1 \times \cdots \times v_n).$$

Here ϕ is **alternating** if ϕ vanishes whenever two of the v_i are equal. A similar property characterizes $S^n E$, with the word "symmetric" replacing "alternating;" ϕ is **symmetric** if $\phi(v_1 \times \cdots \times v_n)$ is independent of the ordering of $v_1, \ldots, v_n$.

For any λ, let $E^{\times \lambda}$ denote the Cartesian product of copies of E, one for each box of the diagram of λ. Consider maps ϕ from $E^{\times \lambda}$ to an R-module F that satisfy:

(1) ϕ is multilinear: if all entries but one are fixed, ϕ is R-linear in the entries of that box;

(2) ϕ is alternating in the columns: if $\mathbf{v} \in E^{\times \lambda}$ has two equal entries in the same column, then $\phi(\mathbf{v}) = 0$;

(3) ϕ satisfies the following **exchange property**: choose any two columns, and choose a subset of boxes in the right chosen column. Then for any $\mathbf{v}$ in $E^{\times \lambda}$, $\phi(\mathbf{v}) = \sum \phi(\mathbf{w})$, where the sum is over all $\mathbf{w}$ obtained from $\mathbf{v}$ by interchanging the entries in the chosen boxes in the right column by the entries in the same number of boxes in the left column, maintaining the vertical order in each of the exchanged sets, and keeping the other entries unchanged.

---

[1]There are various names, notations, and constructions in the literature. Some of these make sense only when E is a finitely generated free module, in which case some are dual to others. Some are indexed by the conjugate partition to λ.

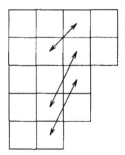

If k boxes are chosen in the right column, and the left column has length c, there are $\binom{c}{k}$ such $\mathbf{w}$ for each $\mathbf{v}$. It suffices, in fact, to do exchanges in adjacent columns, while choosing the top k columns in the right column.

The module E^λ is the universal solution to this problem. This means that there is a map $E^{\times\lambda} \to E^\lambda$, that we can denote $\mathbf{v} \mapsto \mathbf{v}^\lambda$, satisfying (1)–(3), such that for any $\phi\colon E^{\times\lambda} \to F$ satisfying (1)–(3), there is a unique homomorphism of R–modules $\tilde{\phi}\colon E^\lambda \to F$ such that $\tilde{\phi}(\mathbf{v}^\lambda) = \phi(\mathbf{v})$ for all v in $E^{\times\lambda}$.

The module E^λ can be constructed as a quotient module of an appropriate tensor product $E^{\otimes\lambda}$, by dividing by the submodule of relations to make (2) and (3) hold. This construction generalizes the usual construction of symmetric and exterior products.

Recall that one of the main facts about exterior and symmetric powers: if E is free on generators $e_1, \dots, e_m$, then $\wedge^n E$ is free on all $e_{i_1} \wedge \dots \wedge e_{i_n}$, where $i_1 < \dots < i_n$. And $S^n E$ is free on all $e_{i_1} \cdot \dots \cdot e_{i_n}$, where $i_1 \le \dots \le i_n$. The main theorem here gives a corresponding basis for each E^λ. To any filling T of the boxes of λ with integers from $\{1, \dots, m\}$, there is a corresponding element, denoted e_T, of E_λ, that is obtained by replacing each i by e_i, and taking the image under the canonical map from $E^{\times\lambda}$ to E^λ.

THEOREM. *The elements e_T, as T varies over the semistandard tableaux on λ, form a basis of E^λ.*

Recall that a **semistandard tableau** is a filling of λ that is strictly increasing down the columns, and weakly increasing across the rows. It is straightforward, using properties (1)–(3) (and using only adjacent columns and choice of k top boxes for the right column in (3)), to see that these e_T span E^λ. To see that they are independent, take an $m \times n$ matrix $Z_{i,j}$ of indeterminants, with n at least as large as the number of rows of λ, and let $R[Z]$ be the polynomial ring in these indeterminants. For any sequence $K = (k_1, \dots, k_p)$ of at most n integers between 1 and m, let D_K be the determinant of the $p \times p$ matrix $\left[Z_{i,k_j}\right]_{1 \le i,j \le p}$. For any filling T of λ with integers from $\{1, \dots, m\}$, let D_T be the product of the D_K, as K runs over the columns of T. By using an appropriate ordering of monomials in the variables $Z_{i,j}$ (chosen so that the diagonal term of a determinant D_K is smaller than the other terms of the determinant, whenever K is a strictly increasing sequence), it is not hard to see that the D_T, with T semistandard, are linearly independent over R in $R[Z]$. To complete the proof, it suffices to show that there is a mapping from E^λ to $R[T]$ that takes e_T to D_T for each filling T. Property (3) for this mapping follows from a lemma of Sylvester: if A and B are square matrices, and k columns of B are chosen, then $\det(A) \cdot \det(B) = \sum \det(A') \cdot \det(B')$, where the sum is over all matrices A' and B' obtained from A and B by interchanging k columns of A with the given k columns of B, maintaining the ordering in each set of columns. For details, see [F6].

By construction, the group $GL(E)$ acts on E^λ. Each e_T is an eigenvector for the action of the diagonal matrices; the eigenvalue for the matrix $\mathrm{diag}(x_1, \ldots, x_m)$ is the monomial $x^T = \prod x_i^{\#\{i \in T\}}$. The character of this representation is therefore the sum $\sum x^T$, with the sum over all semistandard tableaux on λ with entries in $\{1, \ldots, m\}$. This is one of the standard formulas for the Schur polynomial $s_\lambda(x_1, \ldots, x_m)$.

When $R = \mathbb{C}$, E^λ is the irreducible representation of $GL_n(\mathbb{C})$ of highest weight λ. These give all the irreducible polynomial representations of $GL_n(\mathbb{C})$. The construction of E^λ as a submodule of $\mathbb{C}[Z]$ was discovered by De Ruyts about a century ago. The Jacobi-Trudi representation of the Schur polynomial as a quotient of determinants then amounts to the Weyl character formula for the corresponding representation of $GL_m(\mathbb{C})$. For more on these constructions and their history, see [F6] and its references.

QUANTUM DOUBLE SCHUBERT
POLYNOMIALS (WITH I. CIOCAN-FONTANINE)

Recently there has been some exciting work developing and calculating a deformation of cohomology called quantum cohomology. This works particularly well on homogeneous varieties like flag manifolds. For the complete flag manifold, Fomin, Gelfand, and Postnikov then solved the corresponding "Giambelli problem" of finding formulas for the Schubert varieties in this ring; these are in terms of deformations of Schubert polynomials called quantum Schubert polynomials. In this appendix we give the description of these polynomials from [CF-F], together with some preliminary comments about quantum cohomology and some complementary comments at the end. This involves quantum double polynomials, which were discovered independently (and earlier) by Kirillov and Maeno [K-M], as part of an alternative approach to obtaining the results of [F-G-P].

A homogeneous variety $X = G/P$ has its Chow ring or cohomology ring freely generated by (classes of) Schubert varieties Y_w, where w varies over an index set (the Weyl group when $X = G/B$, or other sets such as partitions when X is a Grassmannian). To each index w there is another $w^\vee$ such that the intersection numbers satisfy $\langle Y_{u^\vee}, Y_v \rangle = \delta_{uv}$ (These intersection numbers are understood to be zero unless the two Schubert varieties have complementary dimensions.) For the classical flag variety $X = Fl(\mathbb{C}^n)$, we have $W = S_n$, and $w^\vee = w_0 \cdot w$.

This means that the product can be written in the form

$$Y_u \cdot Y_v = \sum_w \langle Y_u, Y_v, Y_w \rangle \, Y_{w^\vee},$$

where $\langle Y_u, Y_v, Y_w \rangle$ denotes the triple intersection product, i.e., the number of points in the intersection of the three Schubert varieties, if they are taken in general position; again it is zero if the sum of the codimensions is not the ambient dimension.

To define the (small) quantum cohomology ring, one takes a variable q_i for each $w = s_i$ such that Y_{s_i} has codimension 1 in X. For our case, there are $n - 1$ of these classes. The **quantum cohomology ring** $QH^*(X)$ will be an algebra over the polynomial ring $\mathbb{Z}[q] = \mathbb{Z}[q_1, \ldots, q_{n-1}]$. It is a free algebra over $\mathbb{Z}[q]$ on the same basis $\{Y_w\}$ of Schubert classes, but the multiplication is more interesting. It is given by the formula

$$Y_u \cdot Y_v = \sum_d q^d \sum_w \langle Y_u, Y_v, Y_w \rangle_d \, Y_{w^\vee}.$$

Here the first sum is over all $d = (d_1, \ldots, d_{n-1})$ with each d_i a nonnegative integer, and $q^d = q_1{}^{d_1} \cdots \cdots q_{n-1}{}^{d_{n-1}}$. The coefficients are **Gromov-Witten** numbers, which are numbers of rational curves on X, properly counted. A mapping $f \colon \mathbb{P}^1 \to X$ is said to have degree d if $f_*[\mathbb{P}^1] = \sum_i d_i[Y_{s_i^\vee}]$ in $H_2(X)$. The coefficient $\langle Y_u, Y_v, Y_w \rangle_d$ denotes the number of maps $f \colon \mathbb{P}^1 \to X$ of degree d such that $f(0) \in Y_u$, $f(1) \in Y_v$, $f(\infty) \in Y_w$, where as usual the Schubert varieties are taken to be in general position, and $\langle \ldots \rangle_d$ is defined to be zero unless this number is finite. It takes some work, which we do not discuss here, to show that this definition makes sense, and to prove the remarkable property that this makes $QH^*(X)$ into an *associative* and commutative $\mathbb{Z}[q]$-algebra. (See [F-P] for a discussion of these issues.)

For the case of the complete flag variety, a description of the quantum cohomology ring was given by Giventual and Kim. This was proved, using a model of Bertram for the Grassmannian, as well as the theorem from Chapter 2, by Ciocan-Fontanine [CF], and also by Kim himself. To describe this ring we need the notion of the **quantum elementary symmetric polynomial** $E_k(x_1, \ldots, x_p)$, which is a deformation of the usual elementary symmetric polynomial $e_k(x_1, \ldots, x_p)$. This polynomial can be defined by taking a (Dynkin) diagram with p vertices labeled by $x_1, \ldots, x_p$, with edges between vertex x_i and x_{i+1} labeled q_i. Then $E_k(x_1, \ldots, x_p)$ is the sum of all monomials in $x_1, \ldots, x_p, q_1, \ldots, q_{p-1}$ obtained by choosing vertices and edges that cover exactly k vertices without overlap, and multiplying the corresponding elements together. Note that, when each q_i is given degree two, and each x_i has degree one, $E_k(x_1, \ldots, x_p)$ is a homogeneous polynomial of degree k.

The quantum cohomology ring of the flag manifold is then given by the presentation

$$QH^*(X) = \mathbb{Z}[x_1, \ldots, x_n, q_1, \ldots, q_{n-1}] \big/ \big(E_1(x_1, \ldots, x_n), \ldots, E_n(x_1, \ldots, x_n)\big).$$

Note that one has a canonical additive, but not multiplicative embedding of the usual cohomology $H^*(X)$ into the quantum cohomology $QH^*(X)$, by way of the additive identification $QH^*(X) = H^*(X) \otimes_{\mathbb{Z}} \mathbb{Z}[q]$; we identify c in $H^*(X)$ with $c \otimes 1$ in $QH^*(X)$.

What is needed, in order to use this presentation to calculate the Gromov-Witten numbers, is the quantum analogue of a Giambelli formula, which calculates the class of a Schubert variety in this ring. The solution to this problem was given by Fomin, Gelfand, and Postnikov [F-G-P], who defined quantum Schubert polynomials $\mathfrak{S}_w^q(x)$, and proved that they represent the corresponding Schubert classes Y_w in $QH^*(X)$. For a permutation w in S_n, $\mathfrak{S}_w^q(x)$ is a polynomial in variables $x_1, \ldots, x_{n-1}$ and $q_1, \ldots, q_{n-2}$, homogeneous of degree equal to the length $l(w)$ of w. These quantum Schubert polynomials specialize to the Schubert polynomials $\mathfrak{S}_w(x)$ described in Chapter 1 when each q_i is set equal to 0.

To give their definitions of these polynomials, we need a little more notation. Let

$$\mathcal{J} = \{ J = (j_1, \ldots, j_{n-1}) : 0 \le j_p \le p \text{ for } 1 \le p \le n-1 \},$$

and for $J \in \mathcal{J}$, set

$$e_J(x) = \prod_{p=1}^{n} e_{j_p}(x_1, \ldots, x_p), \qquad E_J(x) = \prod_{p=1}^{n} E_{j_p}(x_1, \ldots, x_p).$$

Write $\mathfrak{S}_w(x) = \sum_{J \in \mathcal{J}} a_J e_J(x)$ for some integers a_J (this is always possible, for unique such integers). Then the **quantum Schubert polynomial** is defined [F-G-P] by

$$\mathfrak{S}_w^q(x) = \sum_{J \in \mathcal{J}} a_J E_J(x).$$

Their proof that this represents Y_w uses the results of [CF], which proved some special cases of this formula, together with some clever combinatorics.

The main purpose of this appendix is to give another description of these polynomials, which is more in the spirit of the classical Schubert polynomials. Recall the description of these polynomials from Chapter 1. Write $w = w_0 \cdot s_{i_1} \cdot \ldots \cdot s_{i_l}$, where $l = l(w_0) - l(w)$. Then

$$\mathfrak{S}_w(x) = \partial_{i_l}^x \circ \cdots \circ \partial_{i_1}^x (\mathfrak{S}_{w_0}),$$

where $\mathfrak{S}_{w_0} = x_1^{n-1} x_2^{n-2} \ldots x_{n-1}$ and ∂_i^x is the operator that takes a polynomial P to $(P - s_i^x(P))/(x_i - x_{i+1})$, with s_i^x the operator that interchanges x_i and x_{i+1}.

There do not appear to be any operators in polynomials in the x's and q's that would give an analogue of this formula for the quantum Schubert polynomials. Nevertheless, we shall prove an equally simple formula for the quantum Schubert polynomials.

The key to this is the observation that the classical Schubert polynomials are specializations of *double* Schubert polynomials, which, as we have seen, have the following definition:

$$(J.1) \qquad \mathfrak{S}_w(x, y) = \partial_{i_l}^x \circ \cdots \circ \partial_{i_1}^x (\mathfrak{S}_{w_0}(x, y)),$$

where $\mathfrak{S}_{w_0}(x, y) = \prod_{i+j \leq n} (x_i - y_j)$. These double Schubert polynomials, however, have an important symmetry property, that $\mathfrak{S}_w(x, y) = (-1)^{l(w)} \mathfrak{S}_{w^{-1}}(y, x)$ (see equation (1.9)). This means that if we write $w = s_{j_l} \cdot \ldots \cdot s_{j_1} \cdot w_0$, with $l = l(w_0) - l(w)$, then

$$(J.2) \qquad \mathfrak{S}_w(x, y) = (-1)^l \partial_{j_l}^y \circ \cdots \circ \partial_{j_1}^y (\mathfrak{S}_{w_0}(x, y)).$$

We will define **quantum double Schubert polynomials** $\mathfrak{S}_w(x, y, q)$ in variables $x_1, \ldots, x_{n-1}, y_1, \ldots, y_{n-1}, q_1, \ldots, q_{n-2}$, that will simultaneously generalize the double Schubert polynomials and the quantum Schubert polynomials, and which are defined by a formula like that of (J.2). Our definition is, with w and w_0 related as in (J.2),

$$(J.3) \qquad \mathfrak{S}_w(x, y, q) = (-1)^l \partial_{j_l}^y \circ \cdots \circ \partial_{j_1}^y (\mathfrak{S}_{w_0}(x, y, q)),$$

with

$$(J.4) \qquad \mathfrak{S}_{w_0}(x, y, q) = \prod_{p=1}^{n-1} E_p(x_1 - y_{n-p}, \ldots, x_p - y_{n-p}).$$

When each q variable is set equal to 0, $E_p(x_1 - y_{n-p}, \ldots, x_p - y_{n-p})$ becomes $\prod_{i=1}^p (x_i - y_{n-p})$, from which we see that

$$(J.5) \qquad \mathfrak{S}_w(x, y, 0) = \mathfrak{S}_w(x, y).$$

Our goal is the

THEOREM. $\mathfrak{S}_w(x,0,q) = \mathfrak{S}_w^q(x)$.

PROOF. It suffices to prove the following claim: each $\mathfrak{S}_w(x,y,q)$ can be written in the form $\sum a_J(y) E_J(x)$ for some polynomials $a_J(y)$ in $\mathbb{Z}[y_1,\ldots,y_{n-1}]$. Granting this claim, we have an equation

$$\mathfrak{S}_w(x,0,q) = \sum a_J(0) E_J(x).$$

Since $\mathfrak{S}_w(x,0,0) = \sum a_J(0) e_J(x)$, and by (J.1), $\mathfrak{S}_w(x,0,0) = \mathfrak{S}_w(x)$, the theorem follows.

To prove the claim, let M be the $\mathbb{Z}[y]$–submodule of the polynomial ring $\mathbb{Z}[x,y,q]$ generated by the $E_J(x)$, as J varies over the index set $\mathcal{J}$. It is obvious that M is preserved by the operators ∂_i^y, so it suffices to prove that $\mathfrak{S}_{w_0}(x,y,q)$ is in M. But this follows from the easy identity

$$E_p(x_1+Y,\ldots,x_p+Y) = \sum_{i=0}^{p} E_i(x_1,\ldots,x_p)Y^{p-i},$$

and our definition of $\mathfrak{S}_{w_0}(x,y,q)$.

A similar argument shows that any $\mathfrak{S}_w(x,y,q)$ can be written in the form $\sum_{u \in S_n} T_u(y)\mathfrak{S}_u^q(x)$. Setting the q's equal to 0 and using (1.10), one finds

$$\mathfrak{S}_w(x,y,q) = \sum_{\substack{v^{-1}u=w \\ l(u)+l(v)=l(w)}} (-1)^{l(v)}\mathfrak{S}_v(y)\mathfrak{S}_u^q(x).$$

Here is the table of the quantum double Schubert polynomials for S_3:

$$\mathfrak{S}_{321}(x,y,q) = (x_1-y_2)((x_1-y_1)(x_2-y_1)+q_1)$$
$$\mathfrak{S}_{231}(x,y,q) = (x_1-y_1)(x_2-y_1)+q_1$$
$$\mathfrak{S}_{312}(x,y,q) = (x_1-y_1)(x_1-y_2)-q_1$$
$$\mathfrak{S}_{213}(x,y,q) = x_1-y_1$$
$$\mathfrak{S}_{132}(x,y,q) = x_1+x_2-y_1-y_2$$
$$\mathfrak{S}_{123}(x,y,q) = 1$$

This way of constructing the double Schubert polynomials is the content of [CF-F]. There is another way, proposed by the second named author of this book, to arrive at these double quantum Schubert polynomials. This approach uses the inner product $\langle\ ,\ \rangle$ defined on the polynomial ring $\mathbb{Z}[x_1,\ldots,x_n]$, with values in the ring of symmetric polynomials in these variables (cf. [M1], Chapter 5]). This inner product has the properties that

$$\langle w_0\mathfrak{S}_u(x), \mathfrak{S}_{v\cdot w_0}(\overline{x})\rangle = \delta_{uv} \quad \text{for} \quad u,v \in S_n;$$
$$\langle e_J(x), w_0\overline{x}^{K'}\rangle = \delta_{JK} \quad \text{for} \quad J,K \in \mathcal{J},$$

where $\bar{x} = (-x_1, \ldots, -x_n)$, and $K' = (n - 1 - k_{n-1}, \ldots, 2 - k_2, 1 - k_1)$ (see [M1, (5.11), (5.13)] and [L-S1]). From these equations follow

$$(\text{J.6}) \qquad \bar{x}^I = \sum_{u \in S_n} \langle w_0 \mathfrak{S}_u(x), \bar{x}^I \rangle \, \mathfrak{S}_{u \cdot w_0}(\bar{x});$$

$$(\text{J.7}) \qquad \mathfrak{S}_u^q(x) = \sum_{J \in \mathcal{J}} \langle \mathfrak{S}_u(x), w_0 \bar{x}^{J'} \rangle \, E_J(x).$$

Recall that we are searching for polynomials $\mathfrak{S}_w(x, y, q)$ that simultaneously generalize the double Schubert polynomials $\mathfrak{S}_w(x, y)$ (for $q = 0$), and the quantum Schubert polynomials $\mathfrak{S}_w^q(x)$ (for $y = 0$). Of course, setting

$$\mathfrak{S}_{w_0}(x, y, q) := \sum_{u \in S_n} \mathfrak{S}_u^q(x) \mathfrak{S}_{u \cdot w_0}(\bar{y}),$$

and then defining, for $w \in S_n$,

$$\mathfrak{S}_w(x, y, q) := (-1)^l \partial_{j_l}^y \circ \ldots \circ \partial_{j_1}^y (\mathfrak{S}_{w_0}(x, y, q))$$

(notation as in (J.2)), we get immediately the wanted family of polynomials. Indeed, by (J.2), $\mathfrak{S}_w(x, y, 0) = \mathfrak{S}_w(x, y)$, and looking at the coefficient of the (unique nonzero) element $\mathfrak{S}_{\mathrm{id}}(\bar{y}) = 1$ in $\mathfrak{S}_w(x, 0, q)$, we see that this last polynomial is equal to $\mathfrak{S}_w^q(x)$.

On the other hand, using (J.6) and (J.7), we get

$$\sum_{u \in S_n} \mathfrak{S}_u^q(x) \, \mathfrak{S}_{u \cdot w_0}(\bar{y}) = \sum_{u \in S_n} \sum_{J \in \mathcal{J}} \langle \mathfrak{S}_u(x), w_0 \bar{x}^{J'} \rangle \, E_J(x) \, \mathfrak{S}_{u \cdot w_0}(\bar{y})$$

$$= \sum_{u \in S_n} \sum_{J \in \mathcal{J}} \langle w_0 \mathfrak{S}_u(y), \bar{y}^{J'} \rangle \, \mathfrak{S}_{u \cdot w_0}(\bar{y}) \, E_J(x) = \sum_{J \in \mathcal{J}} \bar{y}^{J'} \, E_J(x).$$

The last expression $\sum_{J \in \mathcal{J}} \bar{y}^{J'} E_J(x)$ is exactly the top double quantum Schubert polynomial $\mathfrak{S}_{w_0}(x, y, q)$ from (J.4).

In [F7] still more general double Schubert polynomials are defined, that specialize to these quantum double Schubert polynomials, and which are related to each other by formulas like (J.1) and (J.2). These "universal Schubert polynomials," in fact, represent degeneracy loci for maps

$$F_1 \to F_2 \to \cdots \to F_{n-1} \to E_{n-1} \to \ldots E_2 \to E_1$$

of vector bundles, but without the restriction that the maps $F_i \to F_{i+1}$ are injective or the maps $E_i \to E_{i-1}$ surjective.

Givental and Kim [G-K] have worked out the small quantum cohomology ring for other homogeneous varieties, but the corresponding Giambelli formulas have not yet been determined.

BIBLIOGRAPHY

[A-C-G-H] E. Arbarello, M.Cornalba, P.A. Griffiths and J. Harris, *Geometry of Algebraic Curves Vol.I*, Springer-Verlag, 1984.

[B-S-S] M. Beltrametti, M. Schneider, and A. Sommese, *Chern inequalities and spannedness of adjoint bundles*, Proceedings of the Hirzebruch 65 Conference in Algebraic Geometry (Ramat Gan, 1993) (M. Teicher, ed.), vol. 9, Israel Math. Conf. Proc., 1996, pp. 97–109.

[B-G] N. Bergeron and A. Garsia, *Sergeev's formula and the Littlewood-Richardson rule*, Linear and Multilinear Algebra **27** (1990), 79–100.

[B-G-G] I.N. Bernstein, I. M. Gelfand and S. I. Gelfand, *Schubert cells and cohomology of the spaces G/P*, Russian Math. Surveys **28:3** (1973), 1–26.

[B-H] S. Billey and M. Haiman, *Schubert polynomials for the classical groups*, J. Amer. Math. Soc. **8** (1995), 443–482.

[B-J-S] S. C. Billey, W. Jockusch and R. P. Stanley, *Some combinatorial properties of Schubert polynomials*, J. Algebraic Combinatorics **2** (1993), 345–374.

[B-G-S] J.-B. Bost, H. Gillet and C. Soulé, *Heights of projective varieties and positive Green forms*, J. Amer. Math. Soc. **7** (1994), 903–1027.

[Br] M. Brion, *The push-forward and Todd class of flag bundles*, Parameter Spaces (P. Pragacz, ed.), vol. 36, Banach Center Publications, 1996, pp. 45–50.

[B-E] D. Buchsbaum and D. Eisenbud, *Algebra structures for finite free resolutions and some structure theorems for ideals of codimension 3*, Amer. J. Math. **99** (1977), 447–485.

[Ch] C. Chevalley, *Sur les décompositions cellulaires des espaces G/B*, Algebraic Groups and their generalizations, Pennsylvania State University 1991 (W. S. Haboush and B. J. Parshall, eds.) Proc. Symp. Pure Math. **56, Part I**, (1994), AMS, 1–23.

[CF] I. Ciocan-Fontanine, *Quantum cohomology ring of flag varieties*, Internat. Math. Research Notes (1995), 263–279.

[CF-F] I. Ciocan-Fontanine and W. Fulton, *Quantum double Schubert polynomials*, Institut Mittag-Leffler Report No. 6, 1996-97.

[DC-P] C. De Concini and P. Pragacz, *On the class of Brill-Noether loci for Prym varieties*, Math. Ann. **302** (1995), 687–697.

[D-P-S] J.P. Demailly, T. Peternell, and M. Schneider, *Compact complex manifolds with numerically effective tangent bundles*, J. Alg. Geom. **8** (1994), 295–345.

[D1] M. Demazure, *Invariants symétriques entiers des groupes de Weyl et torsion*, Invent. Math. **21** (1973), 287-301.

[D2] M. Demazure, *Désingularization des variétés de Schubert généralisées*, Ann. Sci. École Norm. Sup. **7** (1974), 53–88.

[E-G] D. Edidin and W. Graham, *Characteristic classes and quadric bundles*, Duke Math. J. **78** (1995), 277–299.

[E-L] K. Eriksson and S. Linusson, *Combinatorics of Fulton's essential set*, Duke Math. J. **85** (1996), 61–76.

[F-G-P] S. Fomin, S. Gelfand and A. Postnikov, *Quantum Schubert polynomials*, J. Amer. Math. Soc. **10** (1997), 565–596.

[F-K] S. Fomin and A. N. Kirillov, *Combinatorial B_n-analogues of Schubert polynomials*, Trans. Amer. Math. Soc. **348** (1996), 3591–3620.

[F1] W. Fulton, *Intersection Theory*, Springer-Verlag, 1984, 1998.

[F2]   ———— , *Flags, Schubert polynomials, degeneracy loci, and determinantal formulas*,
            Duke Math. J. **65** (1992), 381– 420.
[F3]   ———— , *Schubert varieties in flag bundles for the classical groups*, Proceedings of the
            Hirzebruch 65 Conference in Algebraic Geometry (Ramat Gan, 1993) (M. Teicher,
            ed.), vol. 9, Israel Math. Conf. Proc., 1996, pp. 241–263.
[F4]   ———— , *Determinantal formulas for orthogonal and symplectic degeneracy loci*, J.
            Diff. Geom. **43** (1996), 276– 290.
[F5]   ———— , *Positive polynomials for filtered ample vector bundles*, Amer. J. Math. **117**
            (1995), 627–633.
[F6]   ———— , *Young Tableaux, with applications to representation theory and geometry*,
            Cambridge University Press, 1997.
[F7]   ———— , *Universal Schubert polynomials*, to appear in Duke Math. J..
[F8]   ———— , *Introduction to toric varieties*, Princeton University Press, 1993.
[F-Las]   W. Fulton and A. Lascoux, *A Pieri formula in the Grothendieck ring of a flag bundle*,
            Duke Math. J. **76** (1994), 711–729.
[F-L1]   W. Fulton and R. Lazarsfeld, *On the connectedness of degeneracy loci and special
            divisors*, Acta Mathematica **146** (1981), 271–283.
[F-L2]   ———— , *The numerical positivity of ample vector bundles*, Annals of Math. **118** (1983),
            35–60.
[F-M]   W. Fulton and R. MacPherson, *Categorical framework for the study of singular spaces*,
            Mem. Amer. Math. Soc. **243** (1981).
[F-P]   W. Fulton and R. Pandharipande, *Notes on stable maps and quantum cohomology*,
            Algebraic Geometry, Santa Cruz 1995 (J. Kollár et al., eds.), Proc. Symp. Pure Math.
            62, Part 2 (1997), AMS, 45–96.
[vdG]   G. van der Geer, *Cycles on the moduli space of abelian varieties*, preprint (1997).
[G-K]   A. Givental and B. Kim, *Quantum cohomology of flag manifolds and Toda lattices*,
            Comm. Math. Phys. **168** (1995), 609–641.
[Gl-Ho]   H. Glover and B. Holmer, *Endomorphisms of the cohomology rings of finite Grass-
            mann manifolds*, Lecture Notes in Math. **657** (1978), Springer-Verlag, 179–193.
[G-G]   M. Golubitsky and V. Guillemin, *Stable maps and their singularities*, Springer-Verlag,
            1980.
[Gra]   W. Graham, *The class of the diagonal in flag bundles*, J. Differential Geometry **45**
            (1997), 471–487.
[H-T1]   J. Harris and L. W. Tu, *On symmetric and skew-symmetric determinantal varieties*,
            Topology **23** (1984), 71–84.
[H-T2]   ———— , *Chern numbers of kernel and cokernel bundles*, Invent. Math. vol 75 (1984),
            71–84.
[H-T3]   ———— , *The connectedness of symmetric and skew-symmetric degeneracy loci: odd
            ranks*, Topics in Algebra, Banach Center Publications Vol. 26, Part 2 (S. Balcerzyk et
            al., eds.), PWN–Polish Scientific Publishers, Warsaw, 1990, pp. 249–256.
[Hey]   P. Heymans, *Pfaffians and skew-symmetric matrices*, Proc. London. Math. Soc. **19**
            (1969), 730-768.
[Ho]   M. Hoffman, *Endomorphisms of the cohomology of complex Grassmannians*, Trans.
            Amer. Math. Soc. **281** (1984), 745-760.
[H-H]   P.N. Hoffman, J.F. Humphreys, *Projective representations of the symmetric groups*,
            Oxford University Press, 1992.
[Hu]   J.E. Humphreys, *Linear Algebraic Groups*, Springer-Verlag, 1975.
[J]   J.C. Jantzen, *Representations of Algebraic Groups*, Academic Press, 1987.
[J-L-P]   T. Józefiak, A. Lascoux and P. Pragacz, *Classes of determinantal varieties associated
            with symmetric and skew-symmetric matrices*, Math. USSR Izvestija **18** (1982), 575–
            586.
[J-P-W]   T. Józefiak, P. Pragacz and J. Weyman, *Resolutions of determinantal varieties and
            tensor complexes associated with symmetric and antisymmetric matrices*, Astérisque
            87-88 (1981), 109-189.
[K-L]   G. Kempf and D. Laksov, *The determinantal formula of Schubert calculus*, Acta Math.
            132 (1974), 153–162.

[K-M]   A.N. Kirillov and T. Maeno, *Quantum double Schubert polynomials, quantum Schubert polynomials and Vafa-Intriligator formula*, preprint (1996).

[Kl-La]   H. Kleppe and D. Laksov, *The algebra structure and deformation of pfaffian schemes*, J. of Algebra **64** (1980), 167-189.

[Kn]   D. Knuth, *Overlapping Pfaffians*, Electronic J. Combinatorics **3**(2) (1996), 151-163.

[Ku]   A. Kustin, *Pfaffian Identities with applications to free resolutions, DG-algebras, and algebras with straightening law*, Contemporary Math. **159** (1994), 269-192.

[Lk-Sh1] V. Lakshmibai and C. S. Seshadri, *Geometry of G/P II*, Proc. Indian Acad. Sci. **A 87** (1978), 1-54.

[Lk-Sh2] ———, *Standard monomial theory*, Proc. Hyderabad Conf. on Algebraic Groups, 1989 (S. Ramanan, ed.), Manoj Prakashan, Madras, 1991, pp. 279-322.

[La-La-T] D. Laksov, A. Lascoux and A. Thorup, On Giambelli's theorem for complete correlations, Acta Math. **162** (1989), 143-199.

[Las1]   A. Lascoux, *Anneau de Grothendieck de la variété de drapeaux*, The Grothendieck Festschrift, vol. III, Progress in Math., Birkhäuser, 1990, pp. 1-34.

[Las2]   ———, *Notes on Interpolation*, preprint Tianjin University (1996).

[Las3]   ———, *Puissances extérieurs, determinants et cycles de Schubert*, Bull. Soc. Math. France **102** (1974), 161-179.

[L-P]   A. Lascoux and P. Pragacz, *Operator calculus for $\tilde{Q}$-polynomials and Schubert polynomials*, to appear in Adv. Math..

[L-S1]   A. Lascoux and M. P. Schützenberger, *Polynômes de Schubert*, C. R. Acad. Sci. Paris **294** (1982), 447-450.

[L-S2]   ———, *Schubert polynomials and the Littlewood-Richardson rule*, Letters in Math. Physics **10** (1985), 111-124.

[L-S3]   ———, *Formulaire raisonné de fonctions symétriques*, Prepublication L.I.T.P. Université Paris 7, 1985.

[Lay]   F. Laytimi, *On degeneracy loci*, Int. J. of Mathematics **7** (1996), 745-754.

[M1]   I. G. Macdonald, *Notes on Schubert Polynomials*, Département de mathématiques et d'informatique, Université du Québec, Montréal, 1991.

[M2]   I. G. Macdonald, *Symmetric functions and Hall polynomials*, Oxford University Press, Second edition, 1995.

[Mag]   P. Magyar, *Bott-Samelson varieties and configuration spaces*, preprint.

[Ma]   V. Maillot, *Un calcul de Schubert arithmétique*, Duke Math. J. **80** (1995), 195-221.

[M-S]   D. McDuff, D. Salamon, *J-holomorphic curves and quantum cohomology*, vol. 6, University Lecture Series A.M.S, 1994.

[Mu1]   D. Mumford, *Theta Characteristic of an Algebraic Curve*, Ann. Scient. Éc. Norm. Sup. **4** (1971), 181-192.

[Mu2]   ———, *Prym varieties I*, Contributions to Analysis (L.V. Ahlfors et al., eds.), Academic Press, New York, 1974, pp. 325-350.

[P-P]   A. Parusiński and P. Pragacz, *Chern-Schwartz-MacPherson classes and the Euler characteristic of degeneracy loci and special divisors*, J. Amer. Math. Soc. **8** (1995), 793-817.

[P1]   P. Pragacz, *Enumerative geometry of degeneracy loci*, Ann. Sci. École Norm. Sup. **21** (1988), 413-454; see an errata in [P4, p.172].

[P2]   ———, *Cycles of isotropic subspaces and formulas for symmetric degeneracy loci*, Topics in Algebra (S. Balcerzyk et al., eds.), vol. 26, Part 2, Banach Center Publications, 1990, pp. 189-199.

[P3]   ———, *Algebro-geometric applications of Schur S- and Q-polynomials*, Séminaire d'Algèbre Dubreil-Malliavin 1989-1990 (M.-P. Malliavin, ed.), vol. 1478, Springer Lecture Notes in Math., 1991, pp. 130-191; see an errata in [P4, p.173].

[P4]   ———, *Symmetric polynomials and divided differences in formulas of intersection theory*, Parameter Spaces (P. Pragacz, ed.), vol. 36, Banach Center Publications, 1996, pp. 125-177.

[P-R1]   P. Pragacz and J. Ratajski, *Polynomials homologically supported on degeneracy loci*, Annali della Scuolla Norm. Sup. di Pisa **23** (1996), 99-118.

[P-R2]   ———, *Pieri type formula for isotropic Grassmannians; the operator approach*, Manuscripta Math. **79** (1993), 127–151.

[P-R3]   ———, *A Pieri-type theorem for Lagrangian and odd Orthogonal Grassmannians*, J. reine angew. Math. **476** (1996), 143–189.

[P-R4]   ———, *A Pieri-type theorem for even Orthogonal Grassmannians*, Preprint 96-83, Max-Planck-Institut für Mathematik (1996).

[P-R5]   ———, *Formulas for Lagrangian and orthogonal degeneracy loci; $\tilde{Q}$-polynomial approach*, Compositio Math. **107** (1997), 11–87.

[P-T]   P. Pragacz and A. Thorup, *On a Jacobi-Trudi formula for supersymmetric polynomials*, Adv. Math. **95** (1992), 8–17.

[Sch]   I. Schur, *Über die Darstellung der symmetrischen und der alternierenden Gruppe durch gebrochene lineare Substitutionen*, J. reine angew. Math. **139** (1911), 155-250.

[S1]   F. Sottile, *Pieri's formula for flag manifolds and Schubert polynomials*, Ann. Inst. Fourier (Grenoble) **46** (1996), 89–110.

[S2]   ———, *Enumerative geometry for the real Grassmannian of lines in the projective space*, Duke Math. J. **87** (1997), 59-85.

[Sr]   H. Srinivasan, *Decomposition formulas for Pfaffians*, J. of Algebra **163(2)** (1994), 312-334.

[Ta1]   H. Tamvakis, *Arithmetic intersection theory on flag varieties*, University of Chicago Thesis (1997).

[Ta2]   ———, *Schubert calculus on the arithmetic Grassmannian*, preprint (1997).

[T1]   L. W. Tu, *Degeneracy loci*, Proceedings of the International Conference on Algebraic Geometry (Berlin 1985), Teubner Verlag, Leipzig, 1986, pp. 296-305.

[T2]   ———, *The connectedness of symmetric and skew-symmetric degeneracy loci: even ranks*, Trans. Amer. Math. Soc. **313** (1989), 381–392.

[W]   G. Welters, *A theorem of Gieseker-Petri type for Prym varieties*, Ann. Scient. Éc. Norm. Sup. **18** (1985), 671–683.

[VJ-F]   J. Van der Jeugt and V. Fack, *The Pragacz identity and a new algorithm for Littlewood-Richardson coefficients*, Computers Math. Appl. **21** (1991), 39-47.

INDEX OF NOTATION

$\widetilde{Q}_\lambda(E) = \pi_\lambda(c(E))$ $\widetilde{Q}$-polynomial of bundle, 30
$Q_\lambda(X) = \pi_\lambda(c), c = \prod(1 + x_i)/\prod(1 - x_i)$ Schur Q-polynomial, 30
$\Delta(X) = \prod_{i<j}(x_i - x_j)$, 34, 41
$\alpha + \beta = (\alpha_1 + \beta_1, \ldots \alpha_n + \beta_n)$, 34
$D_\lambda, D^{n,m}, D_\lambda^{n,m}$, 35
$F_\lambda(X/Y)$, 35
$s_k(X - Y), s_\lambda(X - Y)$ Schur polynomial in difference of alphabets, 37
$s_\lambda(E - F) = \Delta_\lambda(s(E)/s(F))$ Schur polynomial in difference of bundles, 38
$s_\lambda(E) = \Delta_\lambda(s(E))$ Schur polynomial of bundle, 38
$\Delta_\lambda(E - F) = \Delta_\lambda(c(E)/c(F))$, 38
λ, μ juxtaposition of partitions, 38
$\mathcal{P}_r$ ideal of polynomials universally supported on degeneracy locus, 39
$G^q E$ Grassmann bundle parametrizing rank q quotients, 40, 42, 44, 49, 51
$G_r E$ Grassmann bundle parametrizing rank r subbundles, 41, 43-5, 51, 57
Δ symmetrizing operator (of Lagrange), 40
∂ Jacobi symmetrizer, 41
$\mathcal{I}_r$ ideal in $\mathbb{Z}[c_\bullet, c'_\bullet]$, 43
$\mathcal{T}_r$ ideal generalizing resultant, 46
$\mathcal{SP}(X|Y) = \mathcal{SP}(X) \otimes_{\mathbb{Z}} \mathcal{SP}(Y)$, 47
$Q_\lambda(E) = \pi_\lambda(c(E) \cdot s(E))$, 48
$l(\lambda)$ length of partition, 48
$P_\lambda(E) = 2^{-l(\lambda)} Q_\lambda(E)$, 48
$P_\lambda(X) = 2^{-l(\lambda)} Q_\lambda(X)$ Schur P-polynomial, 48
$\mathcal{P}_r^s, \mathcal{P}_r^{ss}$ ideal of polynomials universally supported on symmetric or
skew-symmetric degeneracy locus, 48
$K(X)$ Grothendieck group or ring of X, 51, 54, 57, 100
$\chi(X)$ (topological) Euler characteristic, 53-5, 57-60, 62, 64
$c(X)$ Chern class of TX, 53-4, 57-9, 64
$D_{\lambda;\mu}^{a;b}$ binomial determinant, 55
$\Psi(r)$, 57
$W_d^r(C)$ Brill-Noether locus, 60
$h(\lambda)$ product of hook lengths of diagram of partition, 62
$D(a_\bullet)$ degeneracy locus associated with isotropic flag, 79
$OG_n(V), OG_n'(V), OG_n''(V)$ orthogonal Grassmannian, 80
$\Omega(a_\bullet)$ Schubert bundle in isotropic Grassmann bundle, 80
δ diagonal embedding, 80, 127
∇ divided difference operator, 85
$\widetilde{P}_\lambda(E)$ $\widetilde{P}$-polynomial of bundle, 88
$\widetilde{P}_\lambda(x_1, \ldots, x_n)$ $\widetilde{P}$-polynomial, 91
V^r Brill-Noether locus in Prym variety, 93
$[V]$ class of subvariety or subscheme, 104
$c_i(E)$ Chern class, 105
$Z(s)$ zero scheme of section, 106
$\widetilde{Y}_w$ Bott-Samelson scheme, 112-4
$\text{Pf}(X)$ Pfaffian, 115
c characteristic map, 118

GENERAL INDEX

addition theorem (linearity formula), 47
ample divisor, 95
ample vector bundle, 97
arithmetic Schubert calculus, 103
barred permutation, 84
Bézout's theorem, 1
bisymmetric polynomial, 6
Borel characteristic map, 120
Borel-Moore homology, 105
Bott-Samelson scheme, 112-4
Brill-Noether loci, 60-2
    in Prym varieties, 93-6
characteristic map, 118-122
Chern class, 105
Chevalley's formula, 122
class of diagonal, 80-2, 87, 127-8
complete isotropic flag, 65
    standard, 65
complete symmetric polynomial, 4, 27
conjugate (or dual) partition, 6
connectedness theorems, 97-8
correspondence, 24, 114
cycle classes, 104-105
degree, 2
diagonal embedding, 80, 127-8
diagram of permutation, 11
divided difference operator, 9-10, 43, 72-6, 85, 113-4, 136, 138
double Schubert polynomial, 9-10, 12, 15, 101-2, 110, 136, 138
duality formula, 6, 38
essential set, 11, 17
factorization formula, 36-8, 44-5
factorization property for P-polynomials, 49
    for $\widetilde{Q}$-polynomials, 83
family, same or opposite, for isotropic subbundles, 71, 79
flag variety, 20
Gauss-Bonnet formula (or Hopf's theorem), 53